BIOTECHNOLOGY OF PLANT TISSUES

BIOTECHNOLOGY OF PLANT TISSUES

By

Dr. P.R. Yadav
Lecturer
Department of Zoology
D.A.V. College
Muzaffarnagar (U.P.)

&

Dr. Rajiv Tyagi
Department of Zoology
M.M. College
Modi Nagar (U.P.)

DISCOVERY PUBLISHING HOUSE
NEW DELHI-110002

Reprinted - 2019

First Published - 2006

ISBN: 978-81-8356-073-3

Biotechnology of Plant Tissues

Published by:

DISCOVERY PUBLISHING HOUSE PVT. LTD.
4383/4B, Ansari Road, Darya Ganj
New Delhi-110 002 (India)
Phone: +91-11-23279245, 43596064-65
Fax: +91-11-23253475
E-mail: discoverypublishinghouse@gmail.com
sales@discoverypublishinggroup.com
web: www.discoverypublishinggroup.com

Printed at:
Infinity Imaging Systems
Delhi

Preface

The present title "Biotechnology of Plant Tissues" has been designed for all those under-graduates, post-graduates and technicians who wish to know and use the principles and techniques of modern biotechnology. This text provides the essential knowledge of the core processes involved in the cultivation of plant cells and tissues enabling the readers to understand the practical application of these techniques. Chapters like preparation of protoplasts, genetic engineering, micro-propagation, and preparation of haploid cell etc. play an important role in plant breeding programmes. These technique are employed to scale-up the production of desired plants.

A wide range of biotechnological methods including tissue culture techniques, cryptoconservation etc. are now utilized. Language used in this text is simple, straight-forward and easy understandable. It is helped that the book will be proved boon to the students involved in the tissue culture techniques.

There can be no claim to originality except in the manner of treatment and much of the information has been obtained from the books and scientific journals available in the different libraries.

The author expresses his thanks to his friends and colleagues whose continue inspirations have initiated him to bring out this book.

The author is painfully aware of the shortcomings, errors and misprints that have crept in, and shall be grateful to receive suggestion for improvement of the next edition from all the readers.

The author expresses his gratitude to Mr. Wasan and staff of M/s Discovery Publishing House for their whole hearted co-operation in the publication of this book.

Authors

CONTENTS

1

INTRODUCTION

For thousands of years Man has been dependent on plants for food, shelter, medicine and many other purposes. Throughout history he has cultivated and selected the best examples of useful plants to make them more suited to his needs. More recently the sciences of agriculture, horticulture and plant breeding have allowed great improvements to the productivity and quality of crops. The latest advances in genetics and the ability to modify the genetic content of plant cells by genetic engineering means that there is now the potential to make improvements much more quickly than by conventional plant breeding. Such rapid improvements to plants could be very important in a world in which the population is expected to double within 60 years. For such improvements to be successful, plants must be manipulated at the cellular level: an understanding of the aims and techniques of plant cell culture is therefore essential. Plant cell cultures are usually described as being *in vitro* since initially they were enclosed in glass vessels. Now, however, this can be taken to mean a sterile, artificial culture environment.

Historical Aspects

Studies on all aspects of plant development and multiplication in whole plants are often complicated by interactions between the various processes that underlie growth and development. It is, therefore, desirable to simplify matters so that controlling influences can be easily identified and studied. This can be done by isolating and culturing parts of plants *in vitro*.

Attempts in the early 1900s to culture the smallest part of a plant that is considered viable (a cell) were unsuccessful. However, interest in plant cell culture was maintained and in the early 1930s roots were grown *in vitro*. Later in the 1930s isolated portions of storage tissue (such as carrot root) were cultured: callus was then derived from such tissue cultures. Callus is a mass of proliferating tissue consisting mainly of one cell type. If a callus is placed in agitated liquid medium it breaks up to form a suspension of single cells and small cell clumps: this is known as a *cell suspension culture*.

Uses of Plant Cell Culture

Cultures of plants and cells *in vitro* have been used for a number of distinct purposes.

Fundamental Biological Studies

These can provide a simple, easily-manipulated system for investigating a range of phenomena, e.g. the use of cell suspension cultures for studies of cell division is now widespread. However, it must be remembered that results gained from such work may be difficult to apply to whole plants.

High Value Biochemicals

Many of these natural plant metabolites are required by the pharmaceutical and cosmetic industries (e.g. *morphine* and *atropine*). There is a lot of interest in the possibility of commercial production of these compounds from large-scale cell culture. It is hoped that the simplicity of cell culture will allow high yields to be produced at low cost. Since environmental conditions can be tightly controlled *in vitro* a constant output might be maintained without seasonal variation.

Plant Multiplication by Micropropagation

It is in this area that the greatest economic benefits have been seen. It was in the 1920s that development began with the culture of orchid seedlings which are difficult to germinate conventionally. Micropropagation now offers a rapid means of vegetative (asexual) multiplication. However it is generally more expensive than conventional techniques, so its use is thus limited to plants that are otherwise difficult to propagate, have a high value, or where speed of propagation is important. Plants such as *Primula* for instance are easy to propagate conventionally, but new cultivars are multiplied by micropropagation because large numbers may be produced for the market more quickly than by other methods.

Cultivars derived from apical meristems (growth zones) are often found to be free of pathogenic bacteria and viruses which may be present in the parent plant from which the meristem was dissected. If these meristems are dissected out and grown up, they can form a very satisfactory way of initiating and maintaining a disease-free stock of a crop such as potato or geranium.

Embryo Rescue

Hybrid embryos (e.g. formed by fusion of gametes from different species) frequently abort naturally because the endosperm (the nutritive tissue surrounding the embryo) is not compatible with the embryo. By excision and culture on a suitable artificial medium, the embryo can be developed and form a plant.

Pollen Culture

The pollen of many species may be induced to develop into entire plants without fertilizing an ovule (which contains an egg-cell). Such plants are usually haploid (that is, they bear only a single set of chromosomes, as in gametes) and as such can be used as pure breeding lines without the need for self-fertilization for many generations. A number of procedures can be used to cause the chromosome set to double up to form a dihaploid.

Production of Plants with Novel Characteristics

In nature, mutant plants can be produced that are unlike others of their species. These may be of commercial value, for example due to new flower colours. The frequency of such variations is often increased unwittingly, by cell culture techniques. Treatment with chemicals or radiation may result in increased rates of mutation which can cause such variations.

Production of Transgenic Plants

Characteristics such as insect-resistance, that could not be achieved by conventional plant breeding methods can be transferred from one plant species to another by means of a vector (for example by the bacterium *Agrobacterium* spp.) or by 'shooting' cells with micro-projectiles coated with DNA (*biolistics*). Cells which then contain both their own and the added DNA are said to be transformed and can be used to produce transgenic plants.

Protoplast Fusion

Protoplasts (plant cells from which the wall has been removed) can be fused to form somatic hybrids. Such fusion products are the result of the union of two or more protoplasts from similar or dissimilar parents. A plant which combines the desirable characteristics of two or more parent species may then be produced from the fusion product.

Whatever their mode of creation, such manipulated and transgenic cells must be multiplied clonally (asexually) and regenerated into intact plants using *in vitro* cell culture technique.

Germplasm Storage

Many important crops originated from parts of the world currently under severe environmental pressure and other plants with useful genetic traits still remain to be exploited. It is therefore desirable to preserve examples of as many types as possible; whilst it would be preferable that this should be done in their natural environments this may prove difficult so *in vitro* culture is an alternative way to preserve these plants. Very small portions of plant tissues can, in some cases, be stored indefinitely at ultra low temperatures (–196°C): this technique is termed *cryopreservation*. Cryopreservation of plant tissue has to be followed by regeneration back to the whole plant.

Plasticity and Totipotency

Plants, due to their long-life cycle and lack of mobility, have developed a greater capacity to survive extreme conditions than have many animals. To tolerate these extreme conditions many processes in plant growth and development can adapt. Photosynthesis and respiration, frequently coupled with changes in stem and leaf morphology, can adapt to extreme conditions. Plants also have, in many cases, the ability to regenerate lost organs (roots etc) and cell division can be induced from almost all plant tissues to form calli (plural of callus). These are examples of physiological, morphological and regenerative plasticity. Thus, processes involved in plant growth and development can be altered, under appropriate conditions. For example in some ornamental plant species—such as *Forsythia*—an exclusive programme of leaf production can be altered to exclusive flower production. *In vitro*, culture systems generally exhibit great *plasticity*; this plasticity allows one type of culture to be induced to form other types—for example embryos may be formed *in vitro* from somatic cells and haploid cells, as well as from the normal zygote and all these, in turn, could develop into whole plants. We will discuss these culture techniques and their inter-relationships later.

It is commonly assumed that plant cells are totipotent, a term which means that each cell maintains the ability to express the total genetic potential of the whole parent plant, given the correct physical and chemical conditions in which to develop. It appears that mature, non-woody differentiated (i.e. specialized) cells do not lose their genetic potential. In many cases totipotency has not been proved, but for practical purposes in plant tissue culture it is assumed to be true.

NEEDS OF CULTURED CELLS

You may have written down quite an extensive list. We can, however, split the list into two: physical factors and chemical requirements.

In order to live, an organism requires an environment with a suitable temperature range. The extracellular pH must not be so extreme as to interfere with the metabolism of the cells. Some degree of regulation of the intracellular pH is possible—it is usually maintained to within 1 or 2 pH units of neutral (pH 7). The osmotic pressure is also important.

All cells require an energy source and a source of carbon for cellular components. Certain other elements are also required for growth, especially nitrogen, phosphorus and potassium. Others include calcium, oxygen, sulphur, magnesium, iron, manganese, cobalt, copper and zinc. Some of these elements, such as cobalt, are only required in minute amounts, whereas others, such as nitrogen, are needed in relatively large quantities. One other vital requirement of all cells is water, the principal biological solvent.

Table 1.1. Some principal physiological functions of the most important elements for plant nutrition

Element	*Function*
Oxyzen	Common in cell components, electron acceptor
Carbon	Common cellular components, forms basic backbone of most biochemicals
Nitrogen	Part of proteins, nucleic acids and coenzymes
Sulphur	Part of some amino acids and some coenzymes
Potassium	Principal inorganic cation
Magnesium	Important enzyme cofactor and part of chlorophyll
Manganese	Important cofactor
Calcium	Important cellular cation and enzyme cofactor
Iron	Part of cytochromes
Cobalt	Part of some vitamins
Copper	Enzyme cofactor
Zinc	Enzyme cofactor
Molybdenum	Enzyme cofactor

In whole plants, energy from sunlight is used via photosynthesis, to synthesize carbohydrate from carbon dioxide. The essential mineral elements are usually taken up by the roots as water-soluble salts.

How Essential Nutrients are Provided in Culture

We have seen that whole plants are able to sustain themselves from water, sunlight, carbon dioxide and simple mineral solutions. However, when cells are cultured *in vitro*, additional and unique requirements need to be considered. Temperature has to be maintained within limits that permit growth, light has to be provided if the cultures are to carry out photosynthesis. Light quality (wavelength) is also important in regulating many aspects of plant development.

Many cell cultures do not photosynthesize, so they must be provided with a source of fixed carbon (i.e. as part of an organic molecule, usually sucrose). The remaining essential elements are provided by the culture medium—a solution of salts designed to provide elements in the required quantities. Often culture media also provide a source of amino acids and vitamins, since cell cultures may not be able to synthesize their own. *Plant growth regulators* (PGRs) usually have to be added to the medium. PGRs are compounds similar in function to hormones, in that they control growth and development. In summary, media for plant cell cultures can be classified as containing six types of components:

1. Elements usually required in large amounts: nitrogen, calcium, magnesium, potassium, phosphorus, sulphur. These are termed *macro-elements* or *macronutrients*;
2. Elements usually required in lesser amounts than the macro-elements. These include: zinc, cobalt, copper, manganese, molybdenum, iodine, boron. These are termed *micro-elements* or *micronutrients*;
3. Iron source (iron sulphate, usually with another compound, a chelator, that allows slow release into the medium);
4. Organic supplements such as vitamins;
5. Carbon source (usually sucrose);
6. Plant growth regulators.

Stock solutions for each of these are usually prepared, mixed and appropriately diluted. The pH is then adjusted to a physiological value, before the medium is autoclaved (heated under pressure) to sterilize it. It is then dispensed into suitable sterile vessels. The medium, which may be liquid or gelled (solidified) with a substance such as agar thus provides the necessary physiological environment for the growth of the cell culture. The vessels that are used to contain the medium can be of many sizes, shapes and materials, including large and small Petri dishes, Erlenmeyer (conical) flasks, polycarbonate boxes and bioreactors. The type of vessel used has a major effect on the way the cultures grow, but for the moment we shall not consider that point further.

Culture Types and Their Establishment

Plant tissue and cell cultures are usually initiated from pieces or parts of whole plants. Such tissues are termed explants and may be organs such as roots and leaves or specific cell types such as pollen or endosperm. The type and condition of the explant (its size and age) and the way in which it is cultured can have a profound effect on the success (or otherwise!) of culture initiation. Explant characteristics known to *influence tissue* culture initiation and development include explant age, origin or *derivation* (position within the whole organ), size and polarity.

Polarity is a term used to indicate the metabolic consequences of stable orientation of a *plant tissue culture* (PTC) or a part of a plant during culture. In a culture in which tissue is not moved, various gradients of metabolites form in it under the influence of gravity and diffusion of nutrients from the medium. As a result, there is the development of a '*top*' side and a '*basal*' side which will often develop into quite different structures. If, on the other hand, the culture is put into a liquid medium and continuously agitated, polarity will not be imposed.

Two other terms need explanation too; *distal* and *proximal*.

Distal refers to the part of the plant which is furthest from the original point of attachment to the rest of the plant; proximal means nearest to the point of attachment. The distal part of a simple leaf or shoot or root is that part which is at the tip of the leaf or top of the shoot or point (apex) of the root. The term proximal refers to the opposite end of a piece of root, shoot or leaf.

These aspects will be covered in greater detail elsewhere, but young tissue still undergoing cell division usually produces tissue and cell cultures more easily than old tissue. Small explants generally exhibit a high mortality rate whilst large explants might be difficult to manipulate. Many organ explants exhibit polarity in the organization of cell division and morphogenesis which occur when cultured *in vitro*.

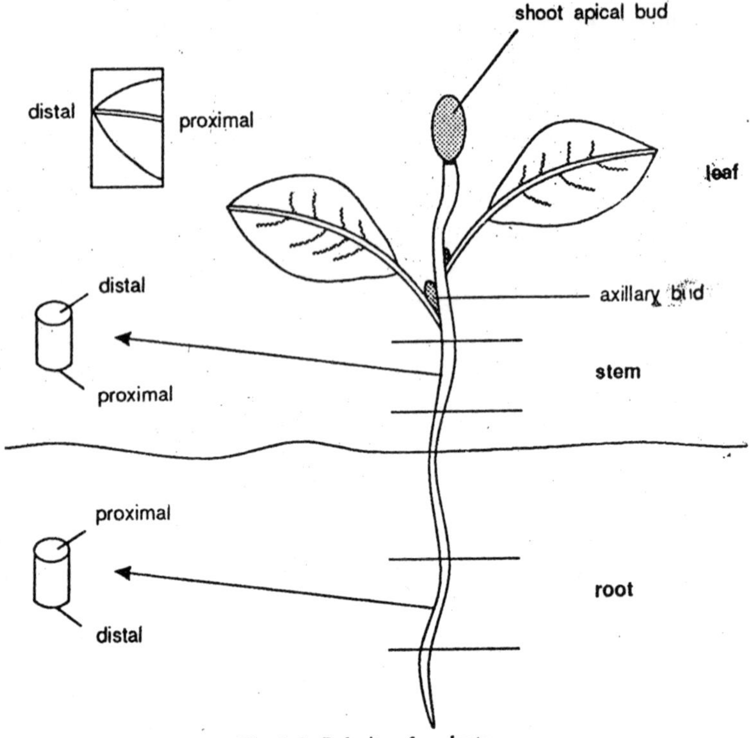

Fig. 1.1. Polarity of explants.

Callus Cultures

On a suitable medium, cultured explants may give rise to a growing, unorganized mass of cells termed a *callus*. It is likely that any nucleated plant cell can, under the right culture conditions, undergo proliferation to produce a callus. During callus formation the processes of cell differentiation (changes in cells during development and specialization) occurring in the whole plant is reversed. This is called dedifferentiation or redifferentiation.

The nutritional requirements of callus cultures frequently include an organic source of reduced nitrogen (amino acids), vitamins and a sugar-alcohol called myoinositol, as well as the usual macro and micronutrients and organic carbon source. In addition, plant growth regulators such as auxins and cytokinins are usually required. Long-term culture of callus tissue may result in the cells loosing the need for a supply of auxin and cytokinin from the medium. Such tissue is said to be *habituated*. Callus tissue is usually cultured on the surface

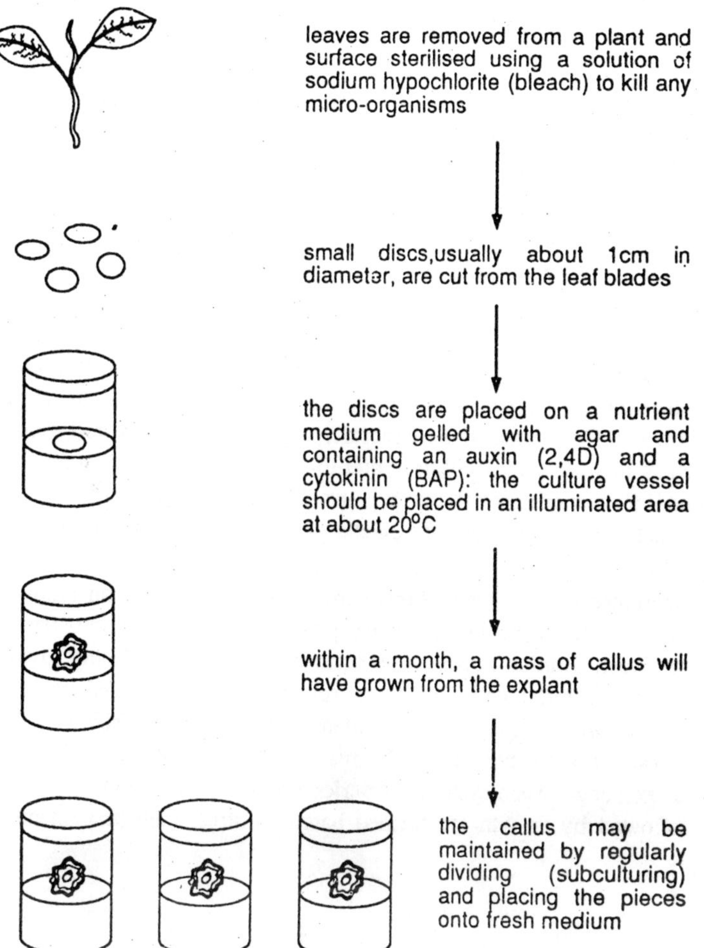

Fig. 1.2. Steps in the production of callus from a pea plant.

of medium that has been solidified and thus only one part (the 'base') of the callus will be in contact with it. This can lead to the generation of various gradients, both in the medium and in the callus. Nutrient, gaseous and toxic waste gradients may all become established. Callus cultures are relatively slow growing, partly due to these gradients, and thus need to be subcultured or transferred, usually every 4 to 6 weeks. Subculture is the process where tissue is periodically moved onto or into fresh medium in order to maintain a supply of nutrients and to prevent the build-up of toxic by-products which would kill the tissue. Callus is usually made up of unspecialized parenchyma cells (thin walled cells). To avoid the necessity for too frequent subculturing, for instance in a germplasm collection, callus may be grown more slowly at cool temperatures (around 10°C). Special media may also be used to slow down the rate of growth.

Cell Suspension Cultures

Some callus cultures become soft and break apart easily: these are said to be 'friable'. When such a friable callus is transferred into agitated liquid medium, usually of the same composition but without the gelling agent, single cells and/or small clumps of cells are released into the medium. These continue to divide and produce a cell suspension. Any large pieces of callus remaining may be removed at the first subculturing. Cell suspensions generally need to be subcultured more frequently than do callus cultures due to their greater growth rate. A sigmoidal (S shaped) growth curve is usually exhibited, much the same as by bacteria in liquid culture. Cells are subcultured at the end of the period of growth.

Subculture of cell suspensions usually involves taking an aliquot (portion) of cells from the stationary phase and diluting it into fresh medium. Such dilution needs to be controlled as cells will only grow if a minimum initial cell density is achieved. Cell suspensions are generally grown in conical flasks which allow a large surface area to volume ratio for the medium to be maintained: this permits efficient gaseous exchange. Such systems of periodic dilution of stationary-phase cells followed by growth are termed batch cultures. Cell suspensions

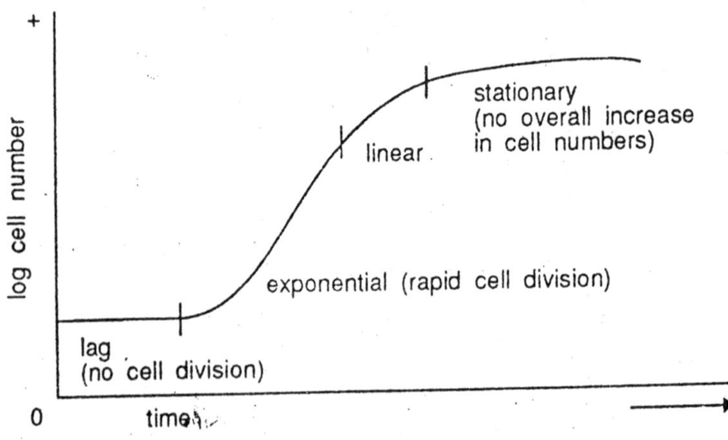

Fig. 1.3. A typical batch growth curve.

can, however, be maintained as continuous cultures in bioreactors where growing cells are constantly diluted with fresh medium and excess cell suspension is removed. Plant cells grown as cell suspensions generally look very much alike from whichever species they originate. Cells of higher plants in suspension rarely have the morphology (form) of cells in the tissue from which they were derived. Cell suspensions may, in some cases be obtained directly from explants, cells growing and being released from the cut surfaces.

The relative homogeneity of cell suspensions has provided biochemists with a system ideally suited to experimentation as many variables can be controlled. They are therefore used as '*model systems*' for studying gene expression, biochemical pathways, the degradation of foreign compounds and as a source of material for enzyme purification.

Protoplasts

Explants, as well as cells in suspension, can also be used as a source of protoplasts. These are plant cells from which the cell wall has been removed, usually by digestion with enzymes such as cellulase or pectinase. Mechanical separation (for instance vortex mixing inside an abrasive tube) alone or in combination with enzymic methods can be used.

During protoplast isolation by enzymatic methods, tissue is incubated in a solution containing a few simple salts, a high sugar concentration and the cell wall degrading enzymes. Isolated protoplasts are recovered, usually by centrifugation, and washed in a similar solution. Protoplasts are again recovered and transferred to a complete culture medium. For efficient protoplast isolation, tissues may need some pretreatment (such as removing the epidermis from leaves) and may have to be of a certain age. The isolation of plant protoplasts from intact plant cells requires a high osmotic potential to prevent them from rupturing and therefore sugars such as sucrose or mannitol in high concentrations are used. This also causes the protoplasm (substance within and including the plasma-membrane) to shrink away from the cell wall during isolation. Protoplasts are fragile and easily damaged either physically or chemically. The handling and culture techniques must therefore be carefully controlled. If they are suspended in liquid medium, it must not be agitated and a high osmotic potential must be temporarily maintained. To provide adequate aeration without agitation, protoplasts are usually cultured in shallow containers. They must also be grown at fairly high cell densities, possibly due to endogenous chemicals tending to leak out and never reaching concentrations essential for normal growth. The culture medium is also often supplemented with additional chemicals and growth factors not needed by most intact plant cells. Protoplasts can divide and produce callus cultures but requires cell wall production. Once cell wall production is initiated the osmotic pressure of the medium can be decreased.

Root Cultures

These can be established *in vitro* from root tips of primary and lateral (side) roots of many plants. As we have already seen, the culturing of roots was the first significant achievement of modem tissue culture. Root cultures can be produced from explants containing primary or lateral root meristems, and will produce a whole root system of primary and lateral roots without other organized structures. The growth of isolated roots is potentially unlimited, roots being 'indeterminate' organs (as distinct from, for example, a leaf which has a set shape and size beyond which it will not grow). Isolated roots can usually be cultured in a fairly simple medium. Liquid medium is usually preferred as growth on gelled medium is slow. However the first major development in modem tissue culture, root culture (at least of untransformed roots) is not widely used.

Shoot Tip and Meristem Culture

The growing tips of shoots may also continue organized growth indefinitely, eventually leading to an organized shoot that can be subsequently rooted; this technique is therefore of great significance in micropropagation.

As discussed already, the culturing of the extreme tip of a shoot (0.2-1 mm), called *meristem culture*, can be used to produce virus-free plants. Larger stem apices' (up to 20 mm) ('shoot tips') cultured *in vitro* are used to micro-propagate many plants. Axillary buds (those formed in the angle between a leaf and the stem) or adventitious buds (not formed from pre-existing buds) can give rise to a clump of shoots which can be rooted. Meristems and shoot tips are usually cultured in standard media, although meristems may require additional vitamins and some amino acids.

Embryo Culture

Embryos (the part of the seed that gives rise to a new plant) can be used as explants in plant tissue culture to produce callus cultures and whole plants. In embryo culture, isolated embryos are used to produce one plant by germination *in vitro*. Seeds or ovules need to be surface-sterilized to remove contaminating micro-organisms but the embryo itself does not need this treatment. The embryo is dissected out of the seed and cultured; mature embryos may need only inorganic salts and sucrose for germination. This technique usually does not lead to rapid and large scale propagation, but is useful if seeds have a long dormant period or if the particular seed genotype has low germination capacity even though the embryo within the seed is viable.

Microspore Culture

Haploid tissues can be produced by culture of immature pollen, (also termed *microspores*), Initially the pollen may be surrounded by somatic tissue (anther). Where anther culture is used, the anthers containing pollen, are removed from the plant and placed on culture medium. Some microspores survive, germinate and some produce haploid plants directly. Special pretreatment of buds or anthers, such as cold temperature treatments, may improve the success of anther culture. When microspores are cultured, agar is avoided, since it has often been found to contain compounds deleterious to pollen growth. Microspores are usually cultured on standard medium and can, given the correct treatments, produce either embryos or calli.

Pollen culture relies on releasing pollen from the anthers prior to cultivation. Pollen (*microspores*) can be released from the anthers by spontaneous release of pollen when the' anthers *dehisce* (open). Anthers can also be homogenized and the homogenate filtered or centrifuged to separate the released microspores. Alternatively, the anther wall can be cut open and the pollen removed—a very time-consuming process but one that avoids cultures being contaminated with diploid tissue and exposure to various harmful compounds released when anthers are homogenized. Haploid cultures are often made to undergo diploidization (doubling of the number of chromosomes per cell) and then used to produce a pure-breeding line of plant.

Plant Regeneration

Plants show a remarkable capacity for changes in the pattern of growth and development due to the plasticity and totipotency of their cells. Plant culture and '*regeneration*' can provide the most startling evidence for totipotency, showing that differentiation does not involve the loss of genetic potential. This capacity for altered development is exploited in plant regeneration.

Plant regeneration can occur via three general methods: *embryo culture*, *somatic embryogenesis* and *organogenesis*. Embryo culture has been dealt with previously in this chapter. Somatic or *asexual embryogenesis* is the production of embryo-like structures from somatic cells. Such embryos, if cultured correctly can develop further and germinate, in a manner analogous to that of *zygotic embryos*. Somatic embryogenesis can occur from various cell types, tissues or organs. The production of somatic embryos can occur either directly or indirectly. The direct mode involves the production of a somatic embryo by a single cell or a group of cells without an intervening callus phase. Direct embryogenesis is common from ovular tissues of *Citrus*, but generally direct embryogenesis is rare in comparison with indirect embryogenesis.

Indirect somatic embryogenesis involves the production of a callus culture from a suitable explant. Embryos are then induced from the callus or from a cell suspension derived from the callus. Classically,

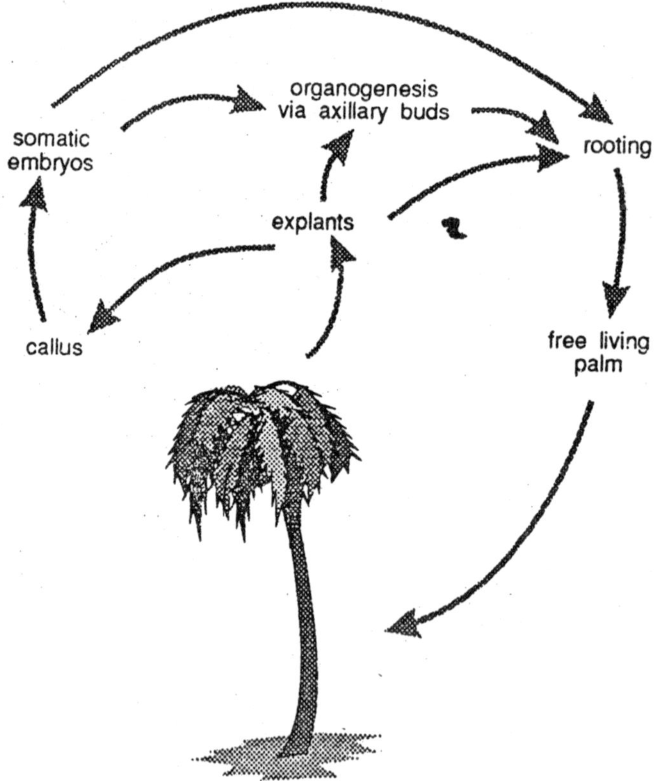

Fig. 1.4. Date palm culture in vitro.

indirect somatic embryogenesis is a two-stage process. Commitment to embryo initiation usually occurs in a medium with a high concentration of the plant growth regulator 2,4-dichlorophenoxy acetic acid. Formation of the embryos then occurs in a medium from which the 2,4-dichlorophenoxy acetic acid has been removed. Indirect somatic embryogenesis has been demonstrated in many plant species. Large numbers of somatic embryos can be generated and this method, although not commercially widespread at the moment, offers an exciting prospect for plant regeneration in the future.

Organogenesis leading to plant regeneration can be achieved through three methods:

(i) via adventitious (arising in abnormal positions) organs forming from a callus, derived from an explant;

(ii) by the formation of adventitious organs directly from explants;

(iii) by the production of plants via axillary bud formation and growth.

Usually organogenesis in cell cultures is controlled principally by the concentrations of various growth regulators present in the culture

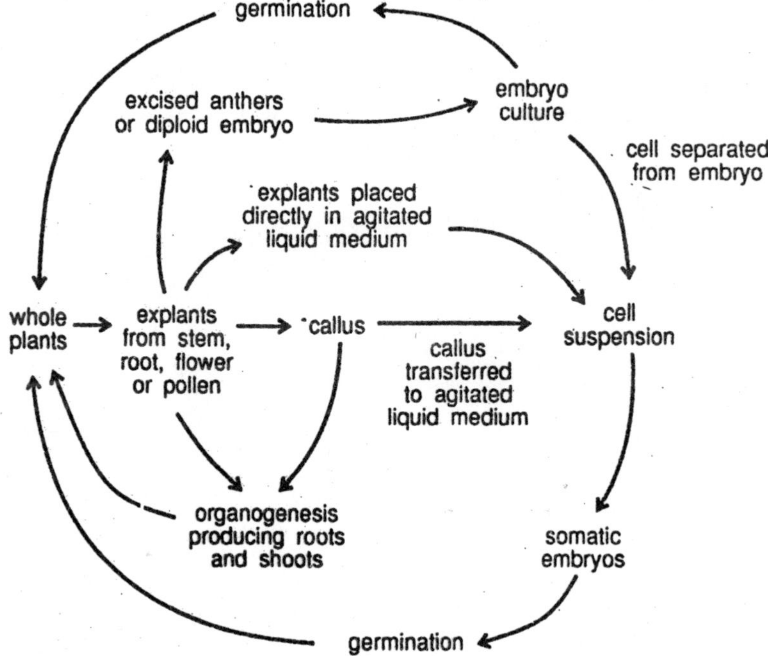

Fig. 1.5. The culture of carrot in vitro.

medium. Tobacco plants provide a good example for organogenesis. Callus can be produced from leaf explants on a medium containing 2,4-dichlorophenoxy acetic acid. If the callus is transferred to medium containing a cytokinin, but without 2,4-dichlorophenoxy acetic acid, shoots are produced from the callus. Shoot development directly from the explants can be achieved by culturing the explants on a medium containing cytokinin as the only plant growth regulator. Shoots derived by either method can be rooted in medium that contains no growth regulators or even in soil. The culture of date palm *in vitro* also illustrates how several types of plant regeneration may be applied to some species. Various explants such as lateral buds, embryos or shoot tips can be cultured and produce plantlets. Regeneration can occur via somatic embryogenesis from the callus produced from the explants. Plantlets and shoots derived from zygotic and somatic embryos, lateral buds and shoot tips can all be induced to form axillary buds. Lateral buds, shoot tips and plantlets can all be rooted on a similar medium, thus free-living plants can be obtained.

It is through the culture of carrot *in vitro* that the inter-relationships of culture and regeneration methods may be demonstrated best. Regeneration of plants from these may be direct, for example via organogenesis or embryo culture, or indirect through a callus stage. Again, notice how many routes for regeneration are possible.

2

FUNDAMENTAL CLUES

Before considering the practical aspects of plant tissue culture, it is important to the aware of the way in which plants grow and develop. Their unique growth form, in which discrete areas of cell division remain active throughout the life of the plant, contrasts markedly with animal development. Understanding the zones of cell division and the way in which their activity is initiated and regulated helps to put into context the principles and methods of plant cell and tissue culture.

Plants show a wide diversity of form and vary in the number and types of organs they possess. The major groups of organs, leaves, roots and stems are present in most plants. Flowers, fruit and storage organs, like tubers, formed at specific stages in the plant's life cycle, also show a variety of forms. Organs may be cultured intact or divided to initiate callus that can differentiate to form organs and embryos. Sections of leaf, for instance, are often used as the host tissue for genetic transformation and embryos are regenerated from the transgenic tissue.

PLANT TISSUES

Plants consist of specialized cell types with differing functions. Plant organs are made of three types of tissue: *dermal tissue* (the epidermis or outer cell layer), *vascular tissue* (the transport system of the plant) and *ground tissues* (all the remaining cells). The type and quantity of each tissue varies for each organ.

Ground Tissues

Ground tissues lie under the epidermis and contribute to structural strength and function. Parenchyma cells are the most abundant cells found throughout the plant and form the bulk of organs like leaves, roots and stems. They have thin flexible cell walls and are initially cuboid, later becoming nearly spherical or cylindrical. In mature tissue, their shape is constrained by surrounding cells. Parenchyma cells contain starch grains and other food reserves and have large vacuoles. Leaf parenchyma cells, termed chlorenchyma, have chloroplasts and are photosynthetic. When the plant is wounded, parenchyma cells divide to form a callus that fills the wound site and may eventually form new tissues to repair the wound. Wound-like callus is frequently formed in plant tissue cultures. Collenchyma cells are similar to parenchyma, but have thickened cell walls and are often found as flexible support beneath the epidermis. Sclerenchyma is a supporting tissue found in organs that have completed lateral growth. It is made up of dead cells with even, lignified secondary cell walls and is organized either as sclereids or as fibres.

Dermal Tissues

All organs are surrounded by dermal tissue that forms a protective layer known as the epidermis. It consists of parenchyma or parenchyma-like cells, and usually forms a complete covering, except where specialized pores (e.g. for gas exchange) are present. The epidermis protects the tissues beneath it from mechanical damage and invasion by pathogens. In roots, the epidermis is involved in nutrient uptake and may possess root hairs.

Vascular Tissue

The vascular tissue made up of xylem and phloem forms the conducting tissues for fluids through the plant. Xylem carries predominantly water and dissolved minerals from root to shoot; phloem carries solutes like sugars and amino acids from sites of synthesis or storage to sites of storage or use.

Callus

Sometimes, a wounded plant will produce a proliferation of undifferentiated cells at the wound site. This mass of cells is known as a *callus*. A similarly disorganized mass of undifferentiated cells produced in tissue culture is also termed *callus*. Callus culture is commonly used as the starting point for various techniques in plant tissue culture, as callus can be induced to differentiate to form organs and embryos.

Meristems

Meristems are regions of actively dividing cells that give rise to cells that differentiate into new tissues of the plant. Apical meristems

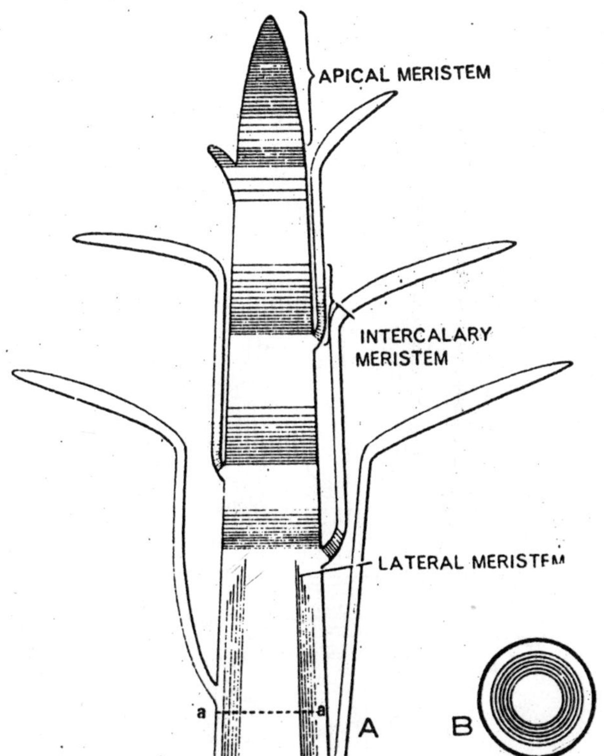

Fig. 2.1. Meristematic tissues. A—Place of meristematic tissues in L.S. of a shoot. B—T.S. of A at a-a.

occur near the tips of roots and shoots and are primarily responsible for length extension of the plant body. This type of growth is called *primary growth*. Meristems contain self-perpetuating cells called *initials* (the equivalent of stem cells in animals), which retain the ability to divide and generate new cells throughout the life of the meristem. Primary meristems that produce the tissues of the stem and root originate in the apical meristems. The primary meristems are the protoderm, the procambium and the ground meristems.

Lateral meristems, the vascular cambium and the cork cambium, produce the secondary tissues and are responsible for increasing the girth of the plant (*secondary growth*). The vascular cambium (or *cambium*) occurs in stem and root, and in mature stems is extended laterally to form a complete cylinder of cells that forms new phloem and xylem. Cork cambium arises in the outer layers of the stems of woody plants typically as a cylinder surrounding the inner tissues.

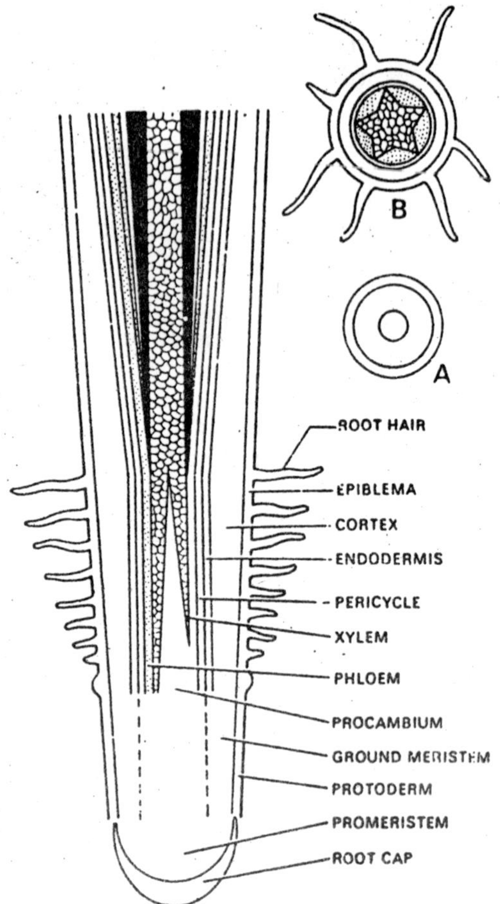

Fig. 2.2. Meristematic tissues. A—Gradual differentiation of tissues in L.S. of root apex. B—Gradual differentiation of tissues in L.S. of shoot apex.

Cells of cork cambium divide to form an outer layer, cork and an inner layer the secondary cortex or phelloderm. Cork, cork cambium and the phelloderm together form an outer protective layer known as the *periderm*.

Meristem culture is used to obtain, maintain and propagate disease-free plants. Viruses, for example, do not invade the meristem and virus-free plants can be produced from cultures of meristematic explants. Meristem culture has been used successfully to eliminate viruses from several species including potato, sugarcane and strawberry.

PLANT REPRODUCTIVE TISSUES

Reproductive tissues are of particular interest because they contain haploid cells with the gametic (*n*) number of chromosomes rather than the (2*n*) found in diploid somatic cells. By the application of cell culture techniques, haploid cells can generate haploid plants. Haploid

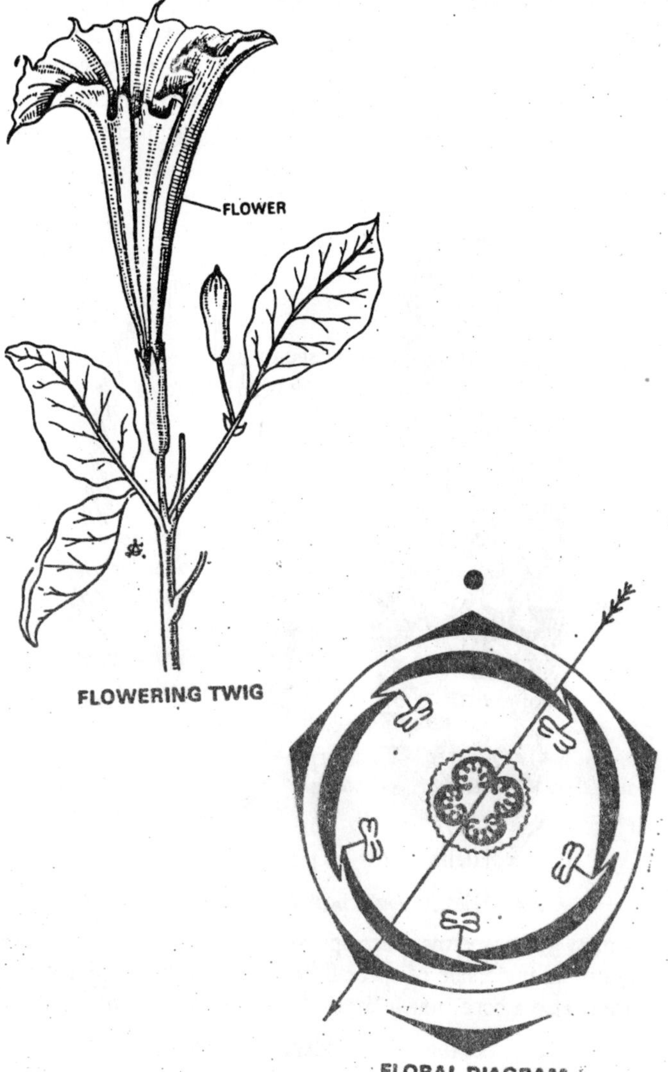

Fig. 2.3. A flowering twig, a flower and floral diagram of Datura metal.

plants can double their chromosome set to form homozygous dihaploid plants. Dihaploids are important elite parents used in breeding programmes for cultivar improvement in several important crops including rice, wheat and *Brassica* species. In angiosperms, the reproductive organ is the flower. *Haploid* pollen (the microspore) is borne in stamens, which comprise an anther supported by a filament. Haploid lines are obtained either by culturing excised anthers or by the culture of isolated microspores. Ovules are more complex and contain an embryo sac in which the haploid embryo nucleus is closely associated with diploid cells. Ovule culture has therefore not been favoured as a means to obtain haploid plants.

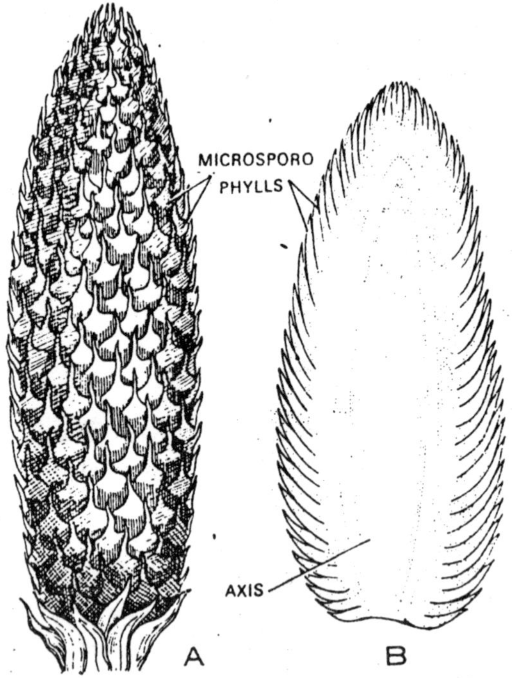

Fig. 2.4. Cycas. A—Entire male cone; B—Male cone in L.S.

In most gymnosperms, microsporangia (anthers) are borne in strobili (cones) on the underside of specialized leaves. The female reproductive structure is also a cone, much larger than the strobili, with woody scales.

Embryo and Embryogenesis

The formation of a plant embryo begins with the zygote, formed from fertilization of an egg cell contained within the embryo sac, by a sperm cell derived from pollen. A second sperm nucleus fertilizes an adjacent cell to form endosperm tissue that will supply nutrients to the developing embryo within the seed.

The basic plan of the plant is established soon alter the ovule is fertilized, in the early stages of the development of the embryo. Embryogenesis in *Arabidopsis thaliana*, a dicot. The fertilized ovule divides to give an apical cell and a basal cell. The basal cell forms the suspensor, connecting the embryo to maternal tissue and the root cap meristem. The apical cell undergoes many cell divisions. The first stage is the octant stage (named from the eight cells formed in two tiers). This is followed by the dermatogen stage (where tangential cell divisions have occurred creating tissue layers). Finally the heart-shaped embryo is formed. This contains the origins of all the major structures of the seedling. The lobes of the heart shape are the cotyledons; between them lies the shoot meristem. The centre of the

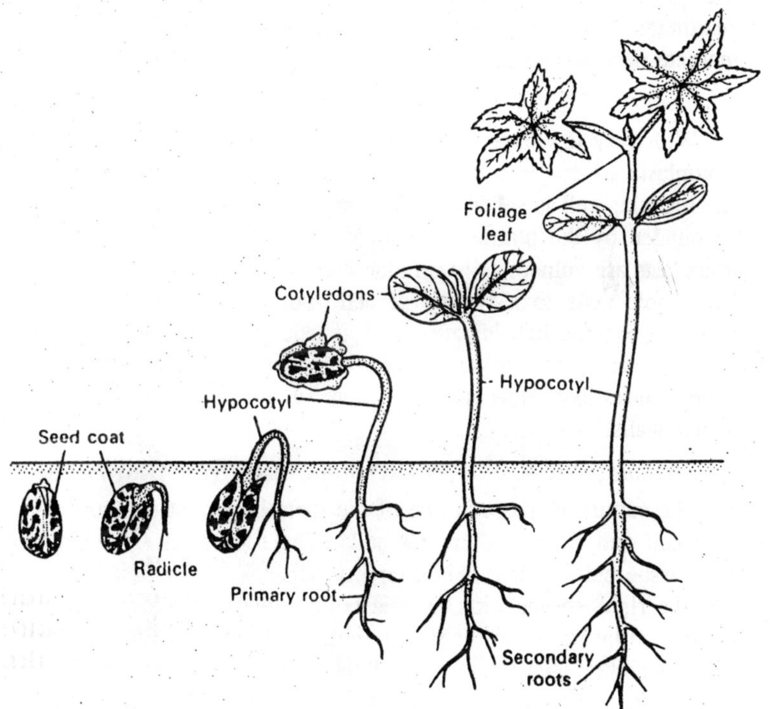

Fig. 2.5. Successive stages of epigeal germination of dicotyledonous and albuminous seed of castor.

heart forms the hypocotyl and the lower layers form the root. The process of embryo formation from zygotes is known as zygotic embryogenesis. Embryos formed in culture from somatic tissue pass through morphological stages of development similar to zygotic embryos. This process is known as *somatic embryogenesis*.

Development of Tissues

The concentric rings of cells laid down in meristems are initially similar in form, non-vacuolate, isodiametric (roughly cuboid) and with thin cell walls. Subsequently they form all the tissues of the plant. This involves major changes in gene expression. The first stage of this is determination, in which the cell becomes established on a pathway of change, but physical changes are not yet detectable. At this stage, the cell is committed to a pathway of development. The cell then becomes differentiated to its new form and function, losing some characteristics and gaining others. In plant cell cultures, both determination and differentiation are frequently reversible given suitable treatments and pathways of determination may be manipulated to generate new tissues and organs, or to induce embryogenesis.

Protoplasts

With very few exceptions, plant cells are enclosed by a cell wall. Protoplasts are single cells that have been stripped of their cell walls. The resulting cell is spherical and consists of the original cell contents surrounded by the plasma membrane. In the absence of a cell wall, protoplasts are vulnerable to osmotic lysis, but they are more amenable than whole cells to a variety of cell and tissue culture techniques, including somatic hybridization and genetic manipulation. Protoplasts can be isolated from a variety of whole-plant tissues and from plant tissue cultures like callus and suspension cultures by enzymatic digestion of cell walls.

Media, Nutrients and Requirements for Growth

The culture medium is one of the most important components of plant cell and tissue culture methods. The successful application of plant tissue culture procedures largely depends on a culture medium with the right composition. Media compositions therefore vary with the particular culture technique being employed. For example, procedures like regeneration of whole plants from cells or tissues require media with different compositions to initiate each phase of the developmental sequence. In addition to its composition, another important function of the culture medium is to provide the right physical environment for cells and tissues to grow. Solid media, for example, perform a function like soil by providing a physical support matrix where tissue explants can maintain contact with air for gaseous exchange or where regenerated plantlets can root. On the other hand, growing suspension cultures in agitated liquid culture enables the cells to maintain maximum exposure to the ingredients of the medium and facilitates gaseous exchange.

All living plant cells require water, nutrients (inorganic mineral elements and organic compounds) and plant growth regulators (hormones) for sustained growth and development. The components of plant tissue culture media generally reflect these requirements and several specific formulations used routinely for culture have similar basic compositions.

Table 2.1. Components of Murashige and Skoog's medium

Component		*Molarity in final medium*
Major inorganic nutrients (macronutrients)	NH_4NO_3	2.06×10^{-2}
	KNO_3	1.88×10^{-2}
	$CaCl_2.2H_2O$	3.00×10^{-3}
	$MgSO_4.7H_2O$	1.50×10^{-3}
	KH_2PO_4	1.25×10^{-3}
Trace elements (micronutrients)	KI	5.00×10^{-6}
	H_3BO_3	1.00×10^{-4}
	$MnS0_4.4H_2O$	9.99×10^{-5}
	$ZnSO_4.7H_2O$	2.99×10^{-5}
	$CuSO_4.5H_2O$	1.00×10^{-7}
	$CoCl_2.6H_2O$	1.00×10^{-7}
Iron	$FeSO_4.7H_2O$	1.00×10^{-4}
	$Na_2EDTA2H_2O$	1.00×10^{-4}
Vitamins	Myo-inositol	4.90×10^{-4}
	Nicotinic acid	4.66×10^{-6}
	Pyridoxine-HCl	2.40×10^{-6}
	Thiamine-HCl	3.00×10^{-7}
Organic nitrogen source	Glycine	3.00×10^{-5}
Carbon source	Sucrose	8.80×10^{-2}

The requirements for plant cell culture may be subdivided into three groups: inorganic (mineral) nutrients, organic nutrients and plant growth regulators.

Inorganic Nutrients

The inorganic nutrients are mineral elements and based on their essential concentrations are usually classified into two groups:

macroelements that are present in large supply (mM concentrations) and microelements or trace elements that are essential only at very low concentrations (μM).

Macroelements

Nitrogen (*N*) Nitrogen is essential for plant life as it is an integral constituent of many important biological compounds, for example, amino acids, proteins, nucleic acids, chlorophyll and many other biomolecules. In plant culture media, inorganic nitrogen is usually supplied in two forms, in an oxidized form as the nitrate anion (NO_3^-) and in a reduced form as the ammonium cation (NH_4^+). The concentration of total inorganic nitrogen is usually 20-40 mM. Most cultures seem to grow best when supplied with both inorganic forms. In some media formulations organic nitrogen is also added as amino acids or protein hydrolysates. These organic forms of nitrogen cannot fully substitute for the inorganic forms.

Sulphur (*S*) Like nitrogen, sulphur is a vital component of proteins in the amino add residues of cysteine and methionine. Both these amino acids are also precursors for other essential sulphur compounds such as vitamins and co-factors. Sulphur is also present in iron-sulphur proteins, which are important redox mediators of chloroplasts and mitochondria. In plant culture media, sulphur is usually supplied at 1-3 mM as the sulphate anion (SO_4^-) in combination, with magnesium as magnesium sulphate.

Phosphorus (*P*) Phosphorus is a constituent of the nucleic acids DNA and RNA and is abundant in the phospholipids of biomembranes. It is also present in energy-rich phosphate esters like ATP and is an important substrate or product in several metabolic reactions. In plant culture media, phosphorus is supplied as the phosphate anion (PO_4^-) as sodium or potassium dihydrogen phosphate at a concentration of 1-3 mM.

Calcium (*Ca*) Calcium is an important modulator of enzyme action and is required for cell wall synthesis and stabilization. Calcium is used in plant tissue culture media at concentrations of 1-3 mM and is added either as the chloride or nitrate salt.

Magnesium (*Mg*) Magnesium acts as the central atom in the chlorophyll molecule, in the aggregation of ribosomal subunits, which is required for protein synthesis and as an essential cofactor for many enzymatic reactions. Magnesium may also play a role in cation-anion balance. In plant cell and tissue culture media, magnesium is usually added as magnesium sulphate in concentrations of 1-3 mM.

Potassium (K) Potassium is the major cation in plants playing several vital roles. Potassium functions in osmoregulation, cation-anion balance and pH stabilization, as an activator for many enzyme-catalysed reactions and is involved in several key steps in the translation process of protein synthesis. Potassium in tissue culture media may vary with the plant species in culture, but is usually supplied at 20-30 mM in combination with nitrate as potassium nitrate.

Microelements

Iron (Fe) Iron is required for chlorophyll synthesis and functions in redox reactions as a constituent of cytochromes and iron-sulphur proteins. Ferric iron (Fe{III}) is practically insoluble at physiological pHs (the solubility constant for $Fe(OH)_3$ is 10^{-39}) and chelators such as ethylenediamine tetraacetic acid (EDTA) are necessary to keep ferric iron in solution. Fe-EDTA is used in culture media to make iron available over a wide pH range.

Boron (B) Boron is a component part of some structural complexes of cell walls and is required for cell division in the apical meristems of roots. In plant culture media, boron is supplied as boric acid.

Cobalt (Co) It is not dear whether cobalt has any direct function in higher plants. However, cobalt is added to plant culture media at around 0.1 μM.

Copper (Cu) The main function of copper is as protein bound copper in redox reactions. Copper is added to culture media at 0.1 μM as copper sulphate. It is toxic at higher concentrations.

Iodine (I) Iodine is not considered as an essential microelement, but is thought to be a beneficial microelement that improves the growth of roots and callus.

Manganese (Mn) Manganese is found in metalloproteins, where it has a structural, role or acts in a redox capacity. Manganese is usually used in culture media at 5-30 μM as manganese sulphate.

Molybdenum (Mo) Molybdenum is a cofactor of a few redox enzymes including nitrate reductase, which is involved in the conversion of nitrate to ammonia. It is usually supplied at 0.1 μM as sodium molybdate.

Zinc (Zn) Zinc is required for the activity of various types of plant enzymes including dehydrogenases and RNA and DNA polymerases. Zinc is added to culture media at 5-30 μM as zinc sulphate.

Organic Nutrients

While green plants are autotrophic (photosynthetic), most culture systems, at least in the early stages, are heterotrophic and require an organic source of carbon and energy. Cultures are often grown with illumination to initiate the development of chloroplasts and photosynthesis. In addition to an organic source of carbon and energy, plant cultures may also require other complex organic molecules for healthy growth.

Sugars

Sugars are added to plant cultures to supply carbon and energy. Sucrose is the most commonly used sugar in plant culture media, but glucose, fructose and sorbitol are used in some formulations. In addition to its metabolic role, sucrose also acts as an osmoticum and together with the inorganic nutrients helps to balance the osmotic potential of the culture medium.

Vitamins and cofactors

Added vitamins are not essential for plant cell and tissue cultures, although vitamin B_1 (thiamine) is considered beneficial for cultures of some species. However, biotin, pantothenic acid, nicotinic acid (niacin), pyridoxine (pyridoxol; vitamin B_6), folic acid, ascorbic acid (vitamin C) and tocopherol (vitamin E) are added to some media. A single addition of yeast extract is sometimes used as a source of vitamins.

The sugar alcohol *myo*-inositol is frequently used in media for monocots, gymnosperms and some dicots. *Myo*-inositol plays a role in cell wall and membrane development.

Complex organic supplements

Complex additions such as banana powder and the liquid endosperm of the coconut (known as *coconut milk* or *coconut water*) are frequently added to some media to improve growth. Although coconut water is thought to provide growth regulators and other organic compounds, it is not known which components of this supplement are effective in improving the growth of cultures. Because of this uncertainty, some experts discourage the use of these undefined additions.

Plant Growth Regulators

There are five main classes of plant growth regulators or plant hormones that co-ordinate growth and differentiation in plants: auxins, cytokinins, gibberellins, abscisic acid and ethylene. Ethylene is principally involved with abscission, floral senescence and fruit ripening and is rarely used in plant tissue cultures. Of the other four classes of plant hormones, auxins and cytokinins are used frequently in plant tissue cultures and gibberellins and abscisic acid are used occasionally.

Auxins

The major auxin in plants is indole-3-acetic acid (IAA). Auxin is synthesized in the stem and root apices and transported along the plant axis. The primary action of auxin is to stimulate the elongation growth of cells. Shoot growth is stimulated by 10^{-6}-10^{-7} M auxin and root elongation by much lower concentrations (10^{-9}-10^{-10} M).

Auxin also stimulates cell division and differentiation and together with cytokinins regulates several developmental processes. For example, auxin induces lateral root formation in stem cuttings and the differentiation of roots and shoots from callus cultures is controlled by the auxin/cytokinin balance. Similarly, apical dominance, the dominant growth of an apical bud been in many plants, is also auxin/cytokinin regulated.

Auxins treatment causes the formation of parthenocarpic (seedless) fruits in some species (e.g. strawberry, tomato, cucumber, pumpkin and citrus). Senescence and abscission of mature leaves, fruits and flowers is inhibited by auxin; however, abscission of young fruits is enhanced by auxin treatment

In plant cell and tissue cultures, auxin together with cytokinin are used to control differentiation and morphogenesis. However, the naturally occurring auxin is seldom used in plant tissue cultures and the synthetic auxins 2,4-dichlorophenoxyacetic acid (2,4-D), 1-naphthaleneacetic acid (NAA), indole-3-butyric acid (IBA) and p-chlorophenoxyacetic acid are the most common alternatives.

The auxin concentration is a critical factor in plant cell and tissue cultures and the optimum concentration varies from species to species. In most cases, synthetic auxins were first identified from herbicide screens (e.g. 2,4-D is a common herbicide) and at higher concentrations may cause severe growth abnormalities or completely inhibit growth.

Cytokinins

Cytokinins promote growth and development, delay senescence and act with auxins to control growth and development. The first cytokinin to be identified was kinetin, which was isolated from samples of degrading herring sperm DNA that caused large increases of cell division in tobacco cell cultures. Subsequently, the naturally occurring cytokinin zeatin was isolated from the kernels of corn (*Zea mays*). Chemically cytokinins are N_6-substituted derivatives of the nitrogenous purine base adenine. The cytokinins most frequently used in tissue culture media include kinetin, benzylaminopurine, zeatin, 6-(γ,γ-dimethylallylamino) purine and adenine.

Gibberellins

In plants, the major action of gibberellins is stimulation of stem elongation and flowering. Gibberellins are also involved in the mobilization of reserves from the endosperm in the initial stages of embryo growth and germination of cereal seeds. Gibberellins are a very large chemical family with over 90 different gibberellins recognized in plants. Each gibberellin is distinguished by a numerical subscript, for example GA_3. In plant tissue cultures, gibberellins are used to induce organogenesis especially adventitious root formation. Only gibberellic acid GA_3, GA_4 and GA_7 are commonly used in plant cultures.

Abscisic acid

In plants, abscisic acid (ABA) is a sesquiterpenoid (15 carbons) primarily involved in water stress responses, induction of storage protein synthesis in seeds and in seed germination. In plant tissue cultures, ABA' is used to enhance somatic embryogenesis.

Support Matrices

In some plant cell and tissue culture systems, a solid or semi-solid matrix is needed to support tissue explants while allowing contact with the medium. Although a liquid medium is useful for many culture systems, a frequently encountered problem is the condition called *hyperhydricity*. Hyperhydricity is a physiological and morphological disorder of plant tissue cultures, where plants or tissue grow poorly, have a high water content and a glassy or brittle appearance. A support matrix may be created from materials such as filter paper, rafts of polypropylene or discs of polyurethane foam. However, the most commonly used support matrix is formed from gelling agents such as agar, agarose, gelatine or gellan gums. Some gelling agents are also known to *induce hyperhydricity*.

Agar

Agar is the most commonly used gel in plant tissue culture. It is a complex polysaccharide extracted from red algae (*Rhodophyceae*) and comprises two fractions, agarose (70%) and agaropectin (30%). Agarose is the gelling fraction and consists of a polymer of alternating D-galactose and 3,6-anhydrogalactose monosaccharide units. Agaropectin is, the non gelling fraction and consists of a polymer of sulphated D-galactose units. Agar melts at about 100°C and gels at around 35°C. Agar is stable, does not react with media constituents and is not normally digested by plant enzymes.

Agarose

In plant tissue cultures, purified agarose (i.e. without agaropectin) is sometimes used instead of agar, particularly in circumstances where a high purity support matrix is required for example in protoplast and anther culture.

Gellan gums

Gellan gums such as Phytagel (Sigma) and Gelrite (Merck) are polysaccharides obtained from the bacterium *Pseudomonas elodea*. These gelling agents are polymers made up of glucose, glucuronic acid and rhamnose units. They produce clear, colourless gels that make visual screening for contamination easy. Gel strength is determined by the concentration of divalent cations (Mg^{2+} and Ca^{2+}). Poor gelling may occur if the cationic concentration is lower than 4 mM or higher than 8 mM.

3

CULTURE MEDIA

The medium used for plant cell culture must provide all of the elements essential for growth, such as nitrogen and phosphorus and that often organic supplements such as vitamins and amino acids are also included. We have provided lists of typical ingredients of culture media in Tables 3.1, 3.2 and 3.3. Look through these lists carefully before reading on.

Table 3.1. The composition of Gamborg's B5 medium. Concentrations are specified on the table

Component	*Concentration in stock (mg l^{-1})*	*Concentration in medium (mg l^{-1})*	*Volume of stock per litre of medium (ml)*
Macronutrients			
KNO_3	50000	2500	
$CaCl_2.2H_2O$	3000	150	
$(NH_4)_2SO_4$	2680	134	50
$MgSO_4.7H_2O$	5000	250	
$NaH_2PO_4.H_2O$	3000	150	
Micronutrients			
KI	30	0.75	
H_3BO_3	120	3	
$MnSO_4.4H_2O$	400	10	
$ZnSO_4.7H_2O$	80	2	25
$Na_2MoO_4.2H_2O$	10	0.25	
$CuSO_4.5H_2O$	1	0.025	
$CoCl_2.6H_2O$	1	0.025	
Iron source			
FeNaEDTA	3670	36.7	10

Vitamins			
Myo-inositol	add as solid	100	10
Pyridoxine-HCl	1000	1	
Thiamine-HCl	10000	10	1
Nicotinic acid	1000	1	
Carbon source			
Sucrose	add as solid	30g l^{-1}	
Adjust pH to 5.5 before autoclaving			

Table 3.2. The composition and preparation of Murashige and Skoog (MS) medium. Concentrations are specified on the table

Components	*Concentration in (stock) (mg l^{-1})*	*Concentration in medium (mg l^{-1})*	*Volume of stock per litre of medium (ml)*
Macronutrients			
NH_4NO_3	33000	1650	
KNO_3	38000	1900	
$CaCl_2.2H_2O$	8800	440	50
$MgSO_4.7H_2O$	7400	370	
KH_2PO_4	3400	170	
Micronutrients			
KI	166	0.83	
H_3BO_3	1240	6.2	
$MnSO_4.4H_4O$	4460	22.3	
$ZnSO_4.7H_2O$	1720	8.6	5
$Na_2MoO_4.2H_2O$	50	0.25	
$CuSO_4$.5H2O	5	0.025	
$CoCl_2.6H_2O$	5	0.025	
Iron source			
$FeSO_4.7H_2O$	5560	27.8	5
$Na_2EDTA.2H_2O$	7460	37.3	
Vitamins			
Myo-inositol	20000	100	
Nicotinic acid	100	0.5	
Pyridoxine-HCl	100	0.5	5
Thiamine-HCl	100	0.5	
Glycine	400	2	
Carbon source			
Sucrose	add as solid	30g l^{-1}	
Adjust pH 5.7-5.8 before autoclaving			

Table 3.3. The composition and preparation of Schenk and Hildebrandt (SH) medium. Concentration are in milligrams per litre.

Component	*Concentration in stock (mg l^{-1})*	*Concentration in medium (mg l^{-1})*	*Volume of stock per litre of medium (ml)*
Macronutrients			
KNO_3	101000	2525	25
$MgSO_4.7H_2O$	24640	36.6	15
$NH_4H_2PO_4$	11500	287.5	25
$CaCl_2.2H_2O$	14680	220.2	15
Micronutrients			
$MnSO_4$	1320	13.2	
H_3BO_3	500	5	
$ZnSO_4.7H_2O$	100	1	
KI	100	1	10
$CuSO_4.4H_2O$	20	0.2	
Na_2MoO_4	10	0.1	
$CoCl_2.6H_2O$	10	0.1	
Iron source			
$FeSO_4.7H_2O$	1500	15	
Na_2EDTA	2000	20	10
Vitamins			
Thiamine-HCl	500	5	
Nicotinic acid	500	5	10
Pyridoxine-HCl	50	0.5	
Myo-inositol		add as solid	1g per litre
Carbon source			
Sucrose		add as solid	30g per litre
Adjust pH to 5.7 before autoclaving			

Plant cell culture media are designed in such a way to be the composition of several widely utilized and successful way. Tables 3.1, 3.2 and 3.3 list the components, the stock solutions and the proportions of each stock solution that are used to make the complete medium. The media listed are Gamborgs B5, Murashige and Skoog (MS) and Schenk and Hildebrandt (SH).

To make a working solution of a plant cell culture medium the stock solutions are mixed in the correct proportions, diluted and brought to the correct pH before being autoclaved. Why do these media contain the compounds they do and why are the compounds that each medium contains so generally similar? These questions can be answered by looking at the elements that are required for growth and development in plants and examining how these are supplied by plant cell culture media. We can examine each stock solution in turn, look at the compounds it contains and what elements these supply and why they are important for the plant cell cultures.

Table 3.4. The element composition of a typical plant

Element	*Concentration as %dry weight*
Carbon (C)	45
Oxyzen (O)	44
Hydrogen (H)	6
Nitrogen (N)	1.6
Potassium (K)	1.0
Calcium (Ca)	0.5
Magnesium (Mg)	0.2
Phosphorous (P)	0.2
Sulphur (S)	0.1
Chlorine (Cl)	0.015
Iron (Fe)	0.01
Boron (B)	0.002
Manganese (Mn)	0.004
Zinc (Zn)	0.0005
Molybdenum (Mo)	0.00007

MACRONUTRIENTS

This is component of plant cell culture media provides, as its name suggests, the elements that are required in large amounts by cultured plant cells. These are usually considered to be carbon, nitrogen, phosphorous, magnesium, potassium, calcium and sulphur.

Carbon

This is usually supplied in the form of carbohydrate. Sucrose is usually the preferred source because it is cheap, readily available, relatively stable to autoclaving and readily assimilated by plant cells. Other carbohydrates may be used but monosaccharides such as glucose are quickly caramalized by autoclaving and are not so convenient to use. Caramilization is a process by which a variety of products are formed from the carbohydrate giving rise to a brown coloured solution.

Nitrogen

Nitrogen is usually provided by either nitrate ions (NO^-_3) or ammonium ions (NO^+_3) or by both. Ions can be supplied from various salts such as potassium nitrate (KNO_3), ammonium di-hydrogen-orthophosphate ($(NH_4)H_2PO_4$), ammonium sulphate ($(NH_4)_2 SO_4$) and ammonium nitrate ($(NH_4)NO_3$). How nitrogen is supplied to plant cell cultures can have a profound effect on the growth and development of the culture. In living organisms, nitrogen is nearly always present in a reduced form, and to be incorporated into, for example, amino acids, the nitrogen must be in a reduced form, such as NH_4^+. The reduction of nitrogen is an energy demanding process, and therefore there is some advantage in using media which supply nitrogen in a reduced form. If nitrogen is supplied as nitrate, which is highly oxidized, it must be reduced before it can be incorporated into biological compounds.

For several reasons ammonium ions are not usually used as the sole source of nitrogen. In general ammonium ions in high concentration are toxic to biological systems. Historically, nitrogen was provided only by the nitrate ion as ammonium ions were considered to be unnecessary or inhibitory. Ammonium ions only became important components of plant cell culture media from the 1950s when it was realized that plant cells are more tolerant to ammonium ions than are their animal counterparts.

Generally, it is only possible to use ammonium ions as the sole source of nitrogen if the pH of the medium can be maintained at a set value by buffering or by adjusting to the set point during the culture period. This is because the uptake of the ammonium ions by plant cells results in the excretion of H^+ ions which cause acidification of the medium.

Media that have high concentrations of ammonium ions can also cause the formation of *vitrescent* (glass-like) shoots during the culture of some plants. The formation of vitrescent cultures can prove to be a problem if shoot tip culture is being used to multiply a particular type of plant. Vitrification seems less prevalent on media containing little or no ammonium.

The choice of nitrogen source can also affect the pH of plant cell culture media. As we have already seen the uptake of ammonium ions causes acidification of the medium, but the uptake of nitrate ions causes the excretion of OH^- ions and thus causes the medium to become more alkaline. The effects of ammonium and nitrate ion uptake can be used as a partial buffering system in plant cell culture media. It remains effective whilst both nitrate and ammonium ions are present in the medium. The buffering depends on the fact that the uptake of ammonium ions is reduced at acidic pH and the uptake of nitrate ions is reduced when the pH of the medium is alkaline.

Phosphorus

Phosphorus is usually supplied as the phosphate ion PO_4^{3-} from salts such as ammonium, sodium or potassium di-hydrogenorthophosphate ($NH_4H_2PO_4$, $Na_2H_2PO_4$ or KH_2PO_4 respectively). Phosphates are readily absorbed by plant cells in an energy demanding process. The concentration of phosphate in plant cell culture media is usually considerably lower than the nitrate/ammonium concentration. High concentrations of phosphate are thought to inhibit growth, presumably by causing the precipitation of other elements, or by the formation of insoluble calcium phosphates. Phosphates playa key role in the transfer of energy within the cell, regulates the activity of many enzymes and is a component of many biological molecules, such as DNA and RNA.

Potassium

Potassium is usually supplied as the chloride, nitrate or orthophosphate salt in plant cell culture media. Potassium is generally the most important cellular cation and has major roles in cellular homeostasis, such as pH regulation and osmotic regulation. Potassium also regulates the activity of some enzymes. Several physiological processes such as phloem transport and changes in stomatal guard cell turgor also rely on potassium and its ability to be transported across plant cell membranes.

Magnesium

Magnesium is usually supplied as magnesium sulphate ($MgSO_4$). It is a component of chlorophyll and is important biochemically as an enzyme co-factor. It is also a structural component of ribosomes. Magnesium can also fulfil some of the pH and osmotic regulatory functions of potassium.

Calcium

Calcium is usually provided as calcium chloride ($CaCl_2$) although calcium nitrate is also used in some media formulations. The concentration of calcium in fresh medium is often several times greater than that found within cell protoplasts due to the active removal of calcium from cells against a concentration gradient. Removal of calcium from the protoplast is important if precipitation of other elements and the formation of calcium phosphate is to be avoided.

Calcium functions as a second messenger (an intra-cellular signal), is an important enzyme co-factor and enzyme regulator. Calcium ions are also involved in establishing the structure and properties of both cell membranes and cell walls.

Sulphur

Sulphur is usually supplied as magnesium sulphate ($MgSO_4.7H_2O$), although ammonium sulphate is sometimes used as an alternative. Many micronutrients are also supplied as sulphates. Sulphur is an important part of some amino acids and as such it is involved in determining protein structure by the formation of disulphide bridges. Sulphur is also an important part of some enzyme co-factors.

Chloride Ions

As well as being the major source of these elements the macronutrient stock also supplies chloride ions (Cl^-) due to the presence of compounds such as calcium chloride. The chloride ion is readily absorbed and transported by plant tissues. Chloride ions may have an important role in osmotic regulation by balancing cations such as magnesium, potassium and sodium and may be required for the evolution of oxygen during photosynthesis.

MICRONUTRIENTS

The micronutrient stock solution provides the elements that are required only in trace amounts for plant growth and development. Micronutrients commonly added to plant cell culture media include manganese (Mn), copper (Cu), cobalt (Co), boron (B), iron (Fe), molybdenum (Mo), zinc (Zn) and iodine (I). Other elements are also included in some media formulations, although it is doubtful that others beyond those specifically mentioned here are essential for most plant cell cultures. Nevertheless both aluminium (Al) and nickel (Ni) are found fairly regularly. Nickel may be of some benefit in special cases where media with high concentrations of urea are used (urea, as will be discussed later, can be used as a source of nitrogen) as it is a component of the enzyme urease which converts urea to ammonia and carbon dioxide.

Iron is the micronutrient present in the largest quantities and is considered the most important micronutrient. Iron is required for the formation of several chlorophyll precursors and is a component of ferredoxins (proteins containing iron) which are important oxidation : reduction reagents. Iron is also an important component of proteins which carry out oxidation and reduction reactions.

Manganese is required to maintain chloroplast ultrastructure and for photosynthesis. Generally it is required in lesser amounts than iron because many cultures *in vitro* are not autotrophic.

Copper and zinc are important components of some types of enzyme, such as oxidases and superoxide dismutase, which help to prevent damage to tissues due to superoxide radicals. Molybdenum, and iron are important in nitrogen metabolism as they are part of the nitrate reductase and nitrogenase enzymes. Cobalt is a component of vitamin B12. Micronutrients also provide some divalent cations, such as cobalt and manganese, that can substitute for magnesium which is required for the activity of both DNA and RNA polymerase. Boron is required for continued cell division and is used in the formation of DNA and RNA bases.

Iodine is not thought to be essential for the growth of plant cell cultures, but is usually included in micronutrient formulations.

Vitamins

Vitamins were originally associated with animal nutrition and are needed for healthy growth and development. Often plants were used as the source of these vitamins. Although plants synthesize many vitamins, the biosynthetic capability of plant cell cultures *in vitro* may be quite different to that of plants *in vivo* (plants grown in their natural ecosystems). For this reason vitamins are included in plant cell culture media.

There are no firm rules as to what vitamins are essential for plant tissue and cell cultures. Many vitamins are added to plant cell culture media formulations simply because they were used previously. The only two vitamins that are considered to be essential are myo-inositol and thiamine. Myo-inositol is usually considered to be a B vitamin and has many and diverse roles in cellular metabolism and physiology. Inositol forms part of the phosphoinositides and phosphatidylinositol which are important factors in processes such as cell division and act as intra-cellular messengers and enzyme activators. Inositol is also involved in the biosynthesis of vitamin C.

Thiamine is also a B vitamin (B1) and is essential for carbohydrate metabolism and biosynthesis. Thiamine has also been found to influence organogenesis in some plant cell cultures.

It can be seen from Tables 3.1, 3.2 and 3.3 that all three media listed contain at least two vitamins in addition to myo-inositol and thiamine. The use of unnecessary vitamins in plant cell culture media can, due to the relatively high cost of vitamins, result in a considerable wasted expense. Such unnecessary additions continue though, particularly the addition of nicotinic acid and pyridoxine-HCl due to their use in highly successful media such as MS.

These are, of course, general rules. Some particular cultures may require the addition of other vitamins, whilst some cultures (for example, some sugar beet cell suspensions) do not require myo-inositol or thiamine for growth and development.

Other Media Components

As we have already seen, nitrogen has to be in a reduced form for incorporation into cellular components such as amino acids. Supplying amino acids and/or urea to plant cell cultures is another way of providing a reduced source of nitrogen which can be easily assimilated. The most commonly used amino acids are arginine, glutamic acid and glutamine (although asparagine, aspartic acid, alanine and proline are also used). Uptake of amino acids from the medium of plant cell cultures causes a drop in the medium pH, presumably due to a mechanism similar to that used for the uptake of ammonium ions.

Not all amino acids will stimulate the growth of a culture, and indeed many have been found to inhibit the growth of some cell cultures. A careful appraisal of the usefulness of each amino acid is needed before it is incorporated regularly into a medium. As well as providing a source of reduced nitrogen, some amino acids have a strong influence on processes such as embryogenesis.

One amino acid, glycine, is often considered to be a vitamin in plant cell culture and is included in the vitamin stock solution of MS medium. Its continued use seems to be based on historical precedents as no definite proof of its essentiality is available.

One way of providing a mix of amino acids, and therefore of reduced nitrogen, is to add casein hydrolysate to the medium. Casein can be hydrolysed either enzymatically or by acids. Casein hydrolysate is also rich in calcium and phosphate. There are also suggestions that casein hydrolysate contains other, unidentified, growth-promoting factors. The use of casein hydrolysate may have disadvantages, as some of the amino acids produced by its hydrolysis may be inhibitory. Also the precise composition of the casein hydrolysate will vary from batch to batch resulting in a loss of reproducibility when making up media from different stock bottles of the casein hydrolysate.

One common component of plant cell culture media that has not yet been dealt with is ethylene-diaminetetraacetic acid (EDTA). This is usually supplied as the disodium salt (Na_2EDTA). This compound is present so as to allow the slow and continuous release of iron into the medium, which it achieves by complexing with the iron. The sodium and iron salt of EDTA (FeNaEDTA) can also be used as the iron is already present in this form. Iron citrate can also be used. Iron added to a medium in the absence of chelating agents may present other problems. In its ferric state (Fe^{3+}), it may combine with hydroxyl ions to form ferric hydroxide which re-arranges into ferric oxide. Thus is insoluble and will precipitate out of solution.

Plant cell culture media are complex mixtures of inorganic and organic compounds designed to provide all the essential elements required for the growth and development of cultures *in vitro*. The main components can be prepared as stock solutions, by dissolving the required amounts of each salt or other compound in water. These stock solutions can be stored at 4°C (with the exception of the vitamin stock which is stored at -20°C because vitamins are easily destroyed). To produce a medium at the final desired concentration for plant cell or tissue growth, these stock solutions must be mixed in the correct proportions with water. Sucrose is then added and dissolved, the medium adjusted to the correct pH and autoclaved. The sequence to make 1 litre of MS medium would be:

1. Add the separate stock solutions to 400 ml water;
2. Add 30g of sucrose;
3. Add water to 900 ml;
4. Adjust the pH to 5.7 (make sure the sucrose is dissolved first);
5. Add water to 1000 ml;
6. Autoclave.

There are also several important points we must now consider. If the sucrose concentration is too high, or the autoclaving time too long, caramelization will occur as the sucrose hydrolyses (breaks down). This results in the loss of carbohydrate and possibly the production of inhibitory substances. The medium may also become brown.

Not all components of media are autoclavable. These unstable components include some amino acids. These compounds have to be sterilized by filtration through a sterile fine-pore (0.22 μm diameter) membrane filter and added to the medium after it has been autoclaved and cooled to about 60°C. You might have noticed that despite the fact that vitamins are labile in many cases, they are autoclaved. In some cases, as a result of this, the effective concentration of a vitamin may be lower than predicted.

You should now understand the basic composition of complex plant cell culture media, and why the various components have been included. Not all media are so complex though, a much simpler medium, that of White. White's medium contains few macronutrients (and nitrate is the only source of nitrogen): both micronutrients and vitamins are absent but have been replaced by yeast extract. There is also a lack of a compound like EDTA to regulate the release of iron into the medium. Nevertheless, White's medium is a long-established medium that is successful for culturing many tissues, particularly organized structures, such as tomato roots.

Table 3.5. The composition of White's medium

Component	*Medium Concentration (mg/l)*
$CaNO_3$	142
KNO_3	81
$Mg\ SO_4.7H_2O$	74
KCl	65
KH_2PO_4	12
$Fe_2(SO_4)_3$	2.46
Yeast extracts	100
Sucrose	20000

We have now seen the basic composition of several important media types that are widely used in plant cell culture. It is important to realize though that for specialized purposes more specialized media may be needed. These specialized media are often based on pre-existing medium formulations though, so a knowledge of the basics of plant cell culture media is always useful.

Gelling Agents

The media listed previously will obviously only be liquids, but often in plant cell culture a 'solid' medium is used. To make a 'solid' medium, a gelling agent is added to the liquid medium before autoclaving. Gelling agents are usually polymers that set on cooling after autoclaving.

Agar is the most widely used substance for gelling (often incorrectly termed solidifying) plant (and bacterial) culture media. Agars are extracted from seaweed grown in the wild. It is insoluble in cold water but produces relatively inert gels that melt at around 100°C and solidify at around 45°C. The hardness of the medium produced depends on the concentration of agar used and the pH of the medium. Concentrations of 0.6 to 0.8% (w/v) are usually used. Low pH values inhibit the gelling of the agar. While cheap agar can be suitable for commercial micropropagation, for research it is necessary to use purified forms. These are more likely to give reproducible results as agar may contain ions and macromolecules that can affect plant growth. Agar can be washed prior to use with water or solvents such as acetone. Agarose, produced from agar, is a purer compound which is often essential for protoplast preparations. Agarose is much more expensive than agar.

Any natural product is likely to vary from batch to batch. Greater consistency can be achieved by the use of newer products from the growth of bacteria in bioreactors. *Gellan gums* such as Gelrite or Phytogel are now being used increasingly. They are more expensive per unit of weight than agar, but a concentration of only 0.2% (w/v) can be used. Many of these gels may be melted and poured as normal but they set at rather a high temperature compared with agar. Gellan gums also contain impurities, but in spite of that they have been found to be particularly suited to the growth of some tissues (such as banana callus).

Alginate solutions have the property of setting quickly when mixed with a solution containing many divalent cations (such as calcium) and of returning to a fluid form when these cations are removed. These cations can be removed by the use of chelating agents such as EDTA. The most common use of alginate has thus been in the encapsulation of cells to produce secondary metabolites, or somatic embryos to produce synthetic (somatic) seeds. Polyacrylamide gels (produced at room temperature by a catalyst) have also been used for such purposes, but suffer from the disadvantage that the acrylamide monomer from which the gel is formed can be highly toxic to plant cell cultures.

Gelatin (used at around 10% w/v) produces a gel, but this melts at too Iowa temperature to have widespread use in plant tissue culture. A number of other substances, such as polymers of polypropylene oxide and ethylene oxide, or extracts of apple parenchyma have been investigated but are currently of little importance.

Whatever gelling agent is chosen it must be remembered that it may have effects on the rates of diffusion of molecules (both nutrients and toxic by-products) through the medium. Agar has an effect on the water potential of the medium, the concentration chosen can have effects on the development of leaf stomata and the occurrence of vitrification. Aeration may be poor within gelled media and this may inhibit the growth of roots. To increase aeration, agar can be fragmented by passing it through a syringe after autoclaving. This is time-consuming and other methods of support such as plastic mesh, filter paper, cotton wool plugs, sand or paper pulp should be considered. Some of these alternatives have the advantage that they are re-usable and permit changes to be made to the medium without disturbing the culture.

Plant Growth Regulators (PGRs)

Plants are multicellular organisms in their natural state: complex systems for intercellular communication have evolved and permit normal differentiation of cells to roots, stems, leaves and flowers. These developments are governed principally by the production and distribution of a number of classes of *plant growth regulators* (PGRs, plant growth substances or phytohormones).

There are a number of classes of PGRs which we shall consider in turn: auxins, cytokinins, gibberellins, abscisic acid and ethylene. Many herbicides (weed killers) are synthetic analogues of these naturally occurring chemicals and plant cell cultures are readily modified by the addition of such substances. Here we will look at each type of PGR with particular reference to their use in plant cell culture.

It must be remembered that as well as the compounds supplied in known amounts by the experimenter (exogenously), in many cases the tissues themselves continue to produce PGRs (endogenously). The synthesis of endogenous PGRs may be modified by the PGRs supplied exogenously.

Auxins

Auxins have promoting effects on cell division and cell elongation. Indole-3-acetic acid (IAA) is the most important natural auxin, but it is an unstable molecule which can be destroyed by heat and light. Consequently a range of more stable (that is, they can be autoclaved) chemical analogues are used as a source of auxin for plant cell cultures.

Some of these compounds are mammalian carcinogens (and so must be used with great care): they are also considered to be associated with genetic instability (somaclonal variation) in plant tissue culture.

To avoid this problem, relatively stable conjugates of IAA (such as indole-acetyl-L-alanine and indole-acetyl-glycine) have been used with some success.

Table 3.6. Synthetic substitutes for auxin

Usual abbreviation	*Chemical name*
2,4-D	2,4-dichlorophenoxyacetic acid
IBA	3-indolebutyric acid
NAA	1-naphthylacetic acid
NOA	2-naphthyloxyacetic acid
2,4,5-T	2,4,5-trichlorophenoxyacetic acid
MCPA	2-methyl-4-chlorophenoxyacetic acid
Dicamba	2 -methoxy-3,6-dichlorobenzoic acid
Picloram	4-amino-3,5,6-trichloropicolinic acid

Several compounds have been shown to have directly antagonistic effects to auxins either by inhibiting their transport or by preventing normal recognition of auxin molecules in the cell. These compounds are sometimes called anti-auxins. They include 2, 3, 5-tri-indobenzoic acid (TIBA), p-chlorophenoxyisobutyric acid (PCIB) and 2, 4, 6-trichlorophenoxyacetic acid (2,4,6- T).

The similarity in structure between the two compounds enables them to bind both to the same cellular recognition sites. Only 2, 4, 5-T is sufficiently like IAA to produce an auxin type effect; 2,4, 6-T blocks these sites, preventing IAA and other auxins from having an effect.

Cytokinins

Cytokinins are also considered to be promoters of growth. At least 25 structurally related compounds are important naturally. Of these the most commonly used ones in plant cells culture are 4-hydroxy-3-methyl-trans-2-butenylaminopurine (zeatin) and N^6-(2-isopentyl) adenine (2iP or IPA). These compounds are expensive and the synthetic analogues 6-furfurylaminopurine (kinetin) and 6-benzylaminopurine (BAP or BA) are used more often. All these compounds are based on the purine molecule, but substituted phenylureas (such as 1, 3-diphenylurea) have also been found to have considerable activity as cytokinins. Certain chemical analogues of RNA bases can antagonise cytokinin effects.

Gibberellins

Gibberellins or gibberellic acids (GAs) exist as dozens of variations of the gibbane carbon skeleton. Only 2 or 3 active compounds are available commercially, GA_3 being the most widely used. In whole plants, gibberellins generally increase stem length and promote flowering and fruit set. They are used relatively infrequently *in vitro*. Anti-gibberellins such as ancymidol, which act by blocking gibberellic acid synthesis, are available.

Abscisic Acid (ABA)

ABA is generally regarded as being a plant growth inhibitor. It is also involved in stomatal opening.

Ethylene (C_2H_4)

Ethylene (ethene) has many growth regulating functions and is particularly concerned with fruit ripening and plant senescence.

Ethylene is a gas; as such its concentration in the atmosphere surrounding plant cell cultures is affected by the tightness of vessel closures and other factors that are difficult to control. Ethylene may be supplied exogenously as the gas, although the use of 2-chloroethyl-phosphoric acid (ethephon) is more convenient. This compound releases ethylene as it is metabolized in the plant.

Ethylene synthesis may be inhibited by the inclusion into the medium of aminoethoxyvinylglycine (AVG); silver ions antagonise ethylene action. Other volatile substances such as methane, ethane, acetaldehyde and ethanol, which are found in the atmosphere surrounding plant cell cultures may also have growth regulating effects.

Polyamines

The synthesis of polyamines (the main ones being putrescine, spermidine and spermine) is closely linked to that of ethylene. Polyamines are ubiquitous molecules and often have growth regulating activity amongst their many functions. They are not commonly included in plant cell culture media, but have been used to synchronize the development of carrot somatic embryos. In some cases putrescine is thought to stimulate cell division.

Miscellaneious Compounds

Miscellaneous phenolic compounds (such as phloroglucinol and catechol) have been ascribed growth regulating properties. They are thought to enhance the growth of callus, the rooting of shoots and shoot proliferation. This may be due to their reducing properties which tend to inhibit the destruction of IAA. However, there is some possibility that the stimulation of growth is due to the inhibition of concealed (endophytic) bacterial contaminants.

Activated Charcoal

Charcoal is activated by treating it with steam. It is not a plant growth regulator itself but is often included in media since it has great absorbing power. It may thus be used to remove toxic metabolites and growth regulators from plant cell cultures. Polyvinylpyrrolidone (PVP) can serve a similar purpose but is soluble, giving a transparent solution. Charcoal forms a black suspension which may conceal infections and make the examination of cultures difficult, it can however enhance rooting by reducing illumination to the submerged parts of the culture.

Incorporation of PGRs into Media

If included, plant growth regulators are usually added to media at concentrations of 0.1 to 100 micromoles per litre. They are not usually directly soluble in water and so are brought into solution by first dissolving them in a small volume (about 1 ml) of ethanol or acid or alkali. This initial solution can then be diluted with water to give a stock solution. Some chemicals, such as diphenylurea, are difficult to dissolve, and may need to be dissolved in dimethyl sulphoxide (DMSO). Solutions made in DMSO can be added directly to sterile media (in small quantities). Many synthetic growth regulators are relatively heat stable and so may be added to the medium before autoclaving. Most naturally occurring ones, however, (such as gibberellins, zeatin and 2iP) must be filter-sterilized. Manufacturers of PGRs usually supply data on stability and solubility.

Uses and Combinations of Plant Growth Regulators and other Media Additives

Plant Growth Regulators and Tissue Culture

A confusing aspect of the description of the applications and effect of plant growth regulators is that great differences in culture response exist between species and cultivars and even between those of the same plants grown under different conditions. A number of principles can, however, be generally applied. The most widely used types of growth regulators *in vitro* are the auxins and cytokinins. They are frequently employed together, although the concentrations and specific chemicals are varied depending on the desired effects.

Callusing of dicotyledonous plants is most readily induced by an approximately equal amount (typically around 10 μmol l^{-1}) of both auxin and cytokinin. Some care is needed though in interpreting results as different synthetic growth regulators have different potencies.

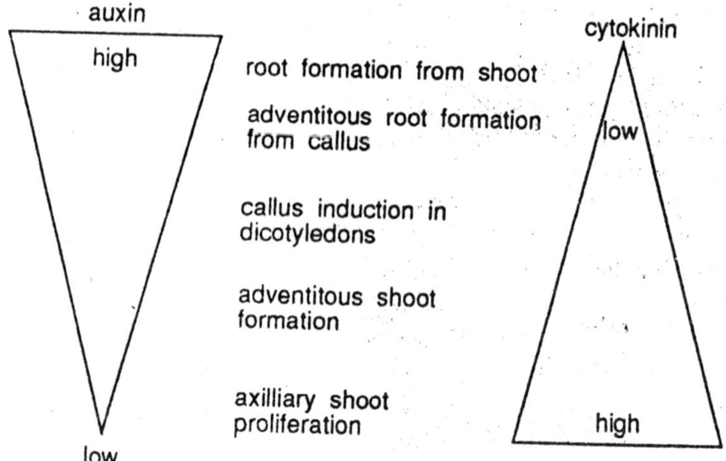

Fig. 3.1. The types of development produced in plant cell cultures by different ratios of auxin to cytokinin.

These compounds have differing stabilities and also the rates of transport and metabolism by plant tissues vary. 2, 4-D is usually considered to be one of the strongest and most commonly used auxins for callus induction and is often used in combination with kinetin for dicotyledons. A similar growth regulator regime may be used for the establishment of cell suspensions from monotyledonous plants but a higher concentration of auxin is normally required. Both callus and cell suspensions of, for example, pearl millet may be initiated and maintained using MS medium supplemented with 2, 4-D at 12 $\mu mol\ l^{-1}$)

Somatic embryogenesis

Two different types of somatic embryogenesis occur. In the direct form embryos are formed in the primary explant, usually in the absence of added growth regulators. However, only certain explants will respond in this manner, these usually being tissues associated with reproduction, such as the nucellus, styles or pollen. Somatic embryos are more commonly formed from callus or cell suspensions subcultured onto secondary medium containing greatly reduced amounts of auxin and cytokinin. Since embryos are not formed on the initial explant, but from an intermediate callus or cell suspension stage, this is termed indirect somatic embryogenesis.

For some species (such as coffee) low concentrations of growth regulators are included in the secondary medium, whilst for others

a) direct

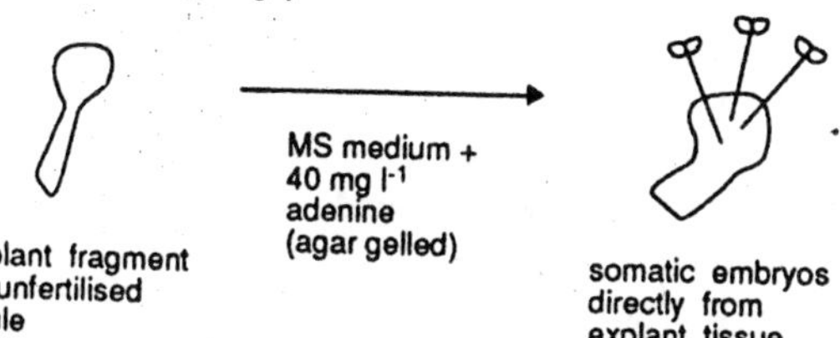

b) indirect

Petunia spp.

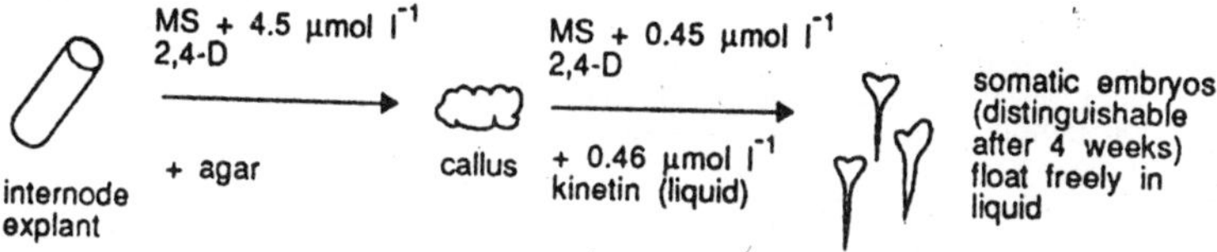

Daucus carota (carrot)

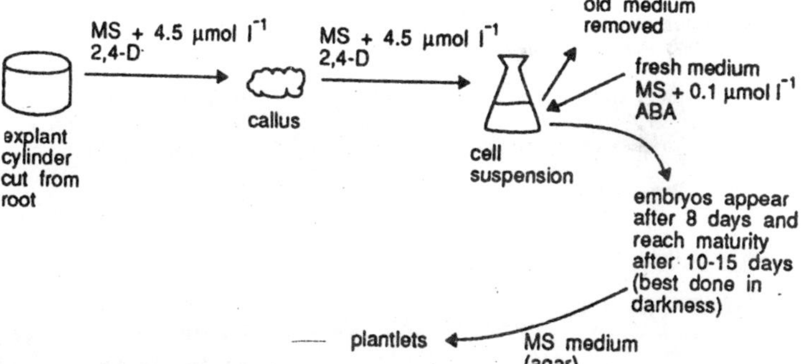

Fig. 3.2. Different types of somatic embryogenesis. (a) Direct embryogenesis; (b) Indirect embryogenesis.

(such as carrot) they may be omitted altogether. PVP or charcoal is sometimes included (for date palms) to absorb any remaining substances carried over on the tissues at subculture. Other substances appear to have subsidiary roles in regulating embryogenesis. The addition of anti-auxins has produced conflicting results. Transport inhibitors such as TIBA may prevent the removal of auxins from the cells following transfer to fresh medium. Gibberellins and ABA can be used to influence embryo germination and dormancy.

Shoots

Shoots may be propagated either through axillary branching or through adventitious shooting. The latter can occur from pre-existing shoot tissue or from callus. It is difficult to induce shoot formation on most roots. Cytokinins are the plant growth regulators mainly associated with shoot proliferation. Generally a high cytokinin to auxin ratio gives the best results. However, the best compounds and concentrations to use vary with the species and conditions used. Axillary bud development of some plants, for example potatoes, can occur readily in the absence of growth regulators.

In some cases gibberellic acid may be used, in place of auxin, plus a cytokinin for shoot induction. The addition of GA_3 to media may also improve the survival of meristem explants, but this response

a) axillary branching (developing from pre-formed buds which although in normal positions, without the growth regulators, may be dormant)

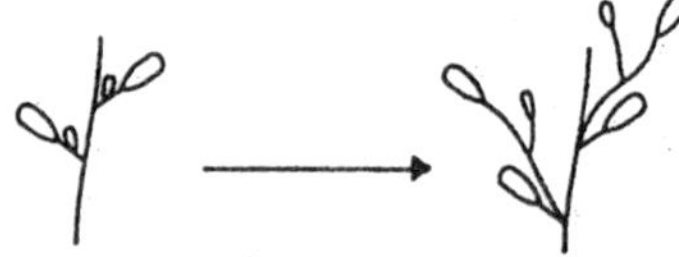

eg *Dracaena spp.*
(0.5 µmol l^{-1} NOA + 23 µmol l^{-1} BAP or kinetin)

eg *Agave spp.*
(1 µmol l^{-1} BAP + 1.7 µmol l^{-1} IAA)

b) adventitious shoot production or caulogenesis (developing from unusual points of origin)

- from callus

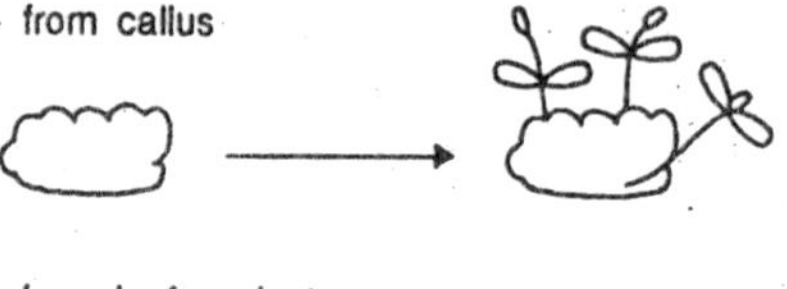

eg *Agave spp.*
(4.6 µmol l^{-1} kinetin + 1.7 µmol l^{-1} IAA)

eg *Freesia spp.*
(23 µmol l^{-1} kinetin)

- from leaf explants

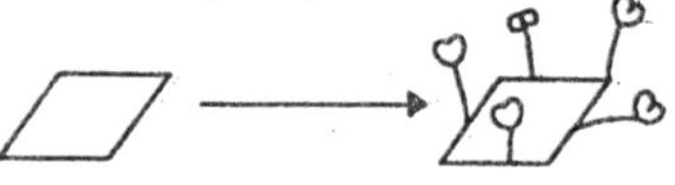

eg African Violet
(5 µmol l^{-1} NAA + 5 µmol l^{-1} BAP)

eg *Begonia spp.*
(1.3 µmol l^{-1} BAP + 5.4 µmol l^{-1} NAA)

- from existing shoots

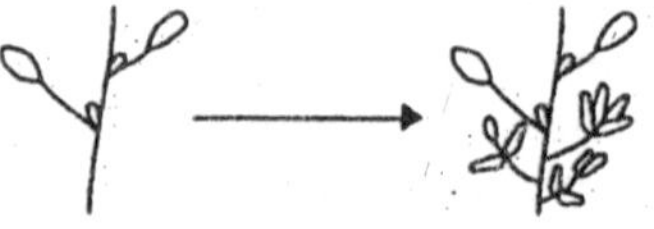

eg *Anthurium spp.*
or (1 to 15 µmol l^{-1} BAP or 5 µmol l^{-1} kinetin)

eg *Gladiolus spp.*
(0.5 µmol l^{-1} NAA 5 µmol l^{-1} kinetin)

Fig. 3.3. Some examples of shoot proliferation induced by specific PGR regimes in the medium. More than one type of shoot can occur at the same time.

is quite variable. One of the few cases in which growth retardants are used in shoot culture is for plants such as *Gladiolius*. Buds in liquid culture may be induced to form small corms.

Roots

The formation of a good root system is a critical part of plant establishment. Roots may be formed during the shoot multiplication phase, particularly if growth regulators are not used. Cytokinins inhibit root formation and so far, the micropropagation of many species depends on the use of a special rooting stage.

The introduction of this extra stage adds to the cost of plant production. By treating the plantlets as ordinary cuttings, roots may be induced non-aseptically *ex vitro* (after *in vitro* culture). A hormone (PGR) (usually an auxin) rooting powder may be used in this case to treat the base of the cut stem.

Adventitious rooting *in vitro* (*rhizogenesis*) is usually stimulated by auxins, usually the weaker synthetic types such as NAA and IBA. 2,4-D, even at low concentrations, tends to result in callusing and has been linked to genetic abnormalities. Although they induce root formation, auxins inhibit root extension. To form *in vitro*, a good root system with some plants such as olive or almond a second, a growth regulator-free medium must be used. This permits the growth of roots induced by an earlier treatment with NAA. Time may be saved by merely dipping the shoots to be rooted in a concentrated auxin solution for a short time prior to planting.

Other Media Additions

The complexity of media necessary for the establishment of shoot tip explants is closely related to their size. Isolated meristems require more stringent culture conditions than do sizable buds. A common problem is tissue and medium blackening which often causes death. This blackening often occurs after the isolation of small explants and is due to the release of phenols which are oxidized by enzymes to produce quinone compounds. These can become bound to proteins and inhibit growth. Several approaches are possible:

Removal of phenolic substances

The explants may be repeatedly transferred to fresh medium, leaving behind the exuded compounds. Their diffusion away from the tissue may also be encouraged by the use of liquid rather than solid medium. Absorption onto PVP or activated charcoal may also effectively remove phenolics.

Modification of redox potential

This can be achieved by the addition of reducing agents to the medium or to the washing solutions used after surface sterilization of the explants. Ascorbic acid, citric acid, cysteine HCl and glutathione may be used.

Oxidase enzyme inhibition

Oxidase enzyme inhibition may be accomplished by the short-term omission of copper from the medium or by the use of chelating agents such as EDTA which may bind copper ions. EDTA is normally included to assist in the provision of iron to the culture, but other substances such as diethyldithiocarbamate may be specifically included (usually at about 10-100 mg l^{-1}) to reduce the availability of copper. The activities of enzymes concerned with phenol metabolism are enhanced by light, so explants of species particularly prone to blackening are often cultured in darkness. Casein hydrolysate and yeast extract are often included in media used for culturing small explants.

Casein hydrolysate and yeast extract provide a source of vitamins, amino acids and other biochemicals to the isolated tissues which would normally receive them from the surrounding cells. Orchid propagation media frequently includes banana homogenate or coconut milk in addition to the usual salts. In many cases such additions are used for historical reasons rather than as a necessity!

4

Apparatus for Changing Tissue Culture Media

Many types of tissue cultures require one or more changes of medium between the time of planting and experimental use. We mechanize these changes as much as possible, so as to reduce the amount of labour.

A medium change simply involves removing the spent medium and adding fresh medium. Either or both of these steps might be mechanized, depending on the number of culture units to be changed, the volume of medium per culture, the configuration of the culture vessel, and the lyre of closure. One would prefer a residue of <5% of the volume of spent medium and a volume of fresh medium dispensed within ± 5% of that selected, though these are seldom critical requirements.

The spent medium may be removed by pouring or by aspiration. Pouring is satisfactory if the number of culture units is small, if the unit volume of medium is quite large, or if several culture vessels share a single closure (as with some types of tube racks). For large numbers of culture units, and especially where the vessel has a constricted opening (as with small plastic flasks), aspiration is preferred. The use of a pipetting machine as an aspirator provides control of the timing and volume of intake. This limits unnecessary drawing of room air into the needle or into the culture vessel. We set the stroke volume at approximately three times the volume of fluid in the cultures; this leaves little residual spent medium, yet the total excess aspirated volume is much less than the free air space in the culture vessel.

Apparatus

If a double pipetting machine with a single-cycle attachment (which stops the machine after a single complete pipetting cycle) is used, one can dispense fresh medium into the cultures in the same operation· During the first ~180° of rotation, the machine's left syringe fills with fresh medium from a reservoir, while the right syringe aspirates the spent medium from the culture. During the second half-cycle, the left syringe dispenses the fresh medium into the culture, and the right syringe discards the spent medium. The machine turns itself off at 360°.

This combination of functions has been made possible by a double, 14-gauge needle produced to our specifications (SH-1085) by the CSI. Spent medium is aspirated with the lower needle, through orifice "A" Fresh medium is dispensed from the upper needle, during the other half-cycle, through orifice "B." The free length of this needle, from mounting point to tip, is ~16 cm. This permits it to reach the bottom of any of the culture vessels we use.

An experienced operator can change 200 screw-cap flask cultures (5-ml fluid volume) per hour using this apparatus. Manipulation of Basks and caps is the rate-limiting factor; the machine cycle takes only ~1 second per culture.

Replicate Culture Methods

Contamination has been no problem in practice. The inherent risk of airborne or manual contamination, using this apparatus, is evidently very low. Neither has transfer of contamination from one culture to another on the needle been a problem, apparently because the first culture has not become contaminated. Our 8 years' experience with this apparatus indicates that it provides a rapid, safe, and economical means of changing medium in large numbers of tissue culture units.

5

Sterile Technique

Successful plant tissue culture requires laboratory facilities where aseptic conditions can be established and maintained. Aseptic technique is critical for plant tissue culture practice because the media and the environment in which plant materials are grown, are also ideal conditions for microorganisms to proliferate. Fungal and bacterial contamination are amongst the hardest to deal with, as these organisms rapidly outgrow plant material and change the defined conditions of the growth medium resulting in poor growth or death of the plant culture.

These deleterious effects are brought about by the contaminating organism(s) consuming the culture nutrients, excreting metabolites into the growth, medium or colonizing the plant tissues. The basic aseptic techniques of handling and culturing plant materials were developed over many years and are designed to ensure that no contaminating micro-organisms enter the culture. Although not always essential, a carefully designed tissue culture unit is recommended to preserve sterile conditions and to achieve consistent results.

Basic Laboratory Layout and Equipment

The prime purpose of a tissue culture laboratory is to enable the processing and culture of plant material in a sterile environment. To facilitate this purpose, certain basic facilities are required. These usually include the following:

1. A media preparation and sterilization area;
2. A sterile transfer area;
3. Environmentally controlled incubators or a culture room.

In laboratories where plant transformation procedures will be routinely used, separate microbiological facilities may also be required for bacterial culture.

Media Preparation Area

The media preparation area should ideally be separate from the culture laboratory and should be equipped with the following:

1. Storage space for chemicals, culture vessels and glassware required for the preparation of media;
2. Refrigerator and freezer space to store stock solutions and chemicals;
3. Bench space with a balance, pH meter, water bath, hotplate and magnetic stirrers;
4. A source of ultrapure water to be used for media preparation;
5. An autoclave and a convection oven for the sterilization of media, glassware and instruments.

Sterile Transfer Area

In order to avoid any contamination that may come from open laboratory benches or from airflow, all manipulations involving culture transfers should be done in a sterile transfer area. Within this area, sources of electricity and gas should be available.

Plastic glove box

The simplest type of transfer area that can be used when relatively few transfers are required is a plastic glove box. This transfer unit can be sterilized by ultraviolet (UV) radiation and by spraying or wiping with 70% ethanol.

Laminar airflow cabinet

Most plant tissue culture facilities use a laminar flow cabinet (or hood) to provide an aseptic area for transfer work. These units are available commercially in different sizes and can be placed where required in the culture laboratory.

The laminar flow cabinet provides a constant flow of filter-sterilized air over the work surface. The air is passed first through a dust filter and then through a high-efficiency particulate air (HEPA) filter with a pore size of 0.2 μm, small enough to remove bacteria and fungal spores. The airflow is then directed either downward (a vertical flow system) or outward (a horizontal flow system) over a non-porous work surface. The constant flow of filter-sterilized air prevents non-sterilized air from entering the working area and thus creates a barrier against airborne contaminants.

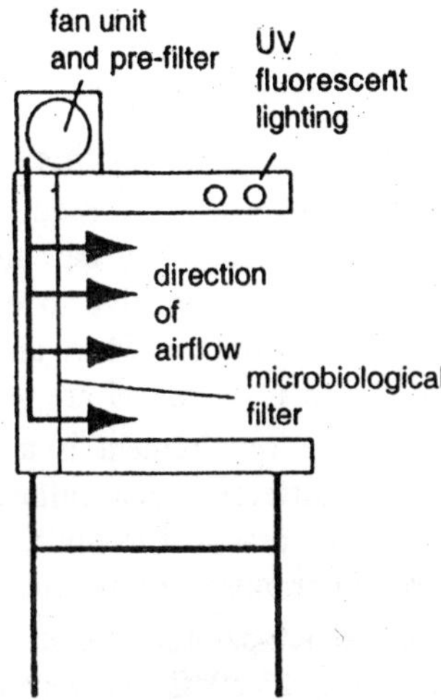

Fig. 5.1. End elevation of a laminar flow cabinet.

The flow cabinet is usually illuminated by fluorescent light and may have a UV light for sterilization of the work surface. The UV light should be switched off when the cabinet is in use. The airflow should remain on continuously or should be run for at least 30 min before using. The work surface should be sterilized before and after use, by swabbing with 70% ethanol. For safety reasons, it is advisable not to use a naked flame like a Bunsen burner in the cabinet as any fires (for instance involving ethanol) are rapidly intensified by the airflow and carried towards the operator. Instead, electrical sterilization devices like a bead or hot-air sterilizer may be used.

Placing of non-sterile materials in the hood should be avoided and as few things as possible should be stored in the cabinet. Any objects opened in the cabinet should be opened facing into the airflow. Before commencing work, hands and forearms should be carefully washed.

In order to minimize contamination, waste material and used apparatus should be removed immediately for sterilization and/or disposal.

Sterile transfer room

In laboratories where large numbers of cultures are routinely processed or large pieces of equipment are required, a useful arrangement would be a dust-free sterile transfer room. This room should have an overhead UV light and a positive-pressure ventilation system equipped with a HEPA filter. It will need gas for a Bunsen burner and bench surfaces should be designed such that they can be easily cleaned and sterilized. The aseptic technique of standard microbiological practice should always be followed.

Culture Room and Plant Growth Facilities

The type of plant growth facilities required would depend on the quantity and type of plant material to be cultured. In general, plant tissue cultures are particularly sensitive to environmental conditions and should be grown under conditions where air circulation, temperaure, humidity, light quality and photoperiod are well controlled.

The minimum culture requirement is a room with temperature control and lighting. If different plant cultures are grown in such a culture unit, it will not be practical to optimize conditions for all the types of culture and compromise settings may have to be adopted.

There is an optimal temperature for each type of plant culture, somewhere in the range 15-30°C. However, for most plants, the temperature is set at 25 ± 2°C. A major problem with a culture room is maintaining a constant temperature throughout the room as hot and cold spots (i.e. temperature gradients) may develop. In addition, the temperature in culture containers may be a few degrees higher than the room temperature because of the greenhouse effect. It is recommended that the room should be fitted with a safety alarm device to indicate when the temperature has reached pre-set maximum or minimum limits.

The culture room should be fitted with fluorescent lighting that can be adjusted for intensity and photoperiod. For mixotrophic cultures, light intensities of approximately 30 μmol m^{-2} s^{-1} photosynthetic active radiation (PAR) are adequate. For autotrophic cultures, light intensities up to 250 μmol m^{-2} s^{-1} PAR will be required. In most cases, a regime of 16 h light and 8 h dark is appropriate. A system where both light and temperature can be programmed for a 24 h period would be advantageous.

Where possible, the culture room should have an adequate ventilation system and the capability of controlling the relative humidity over a range of 20-98%. It should be noted that for cultures grown in standard culture containers, the headspace is usually saturated with water vapour. The relative humidity in the container can be roughly controlled by cooling the base of the container or adjusting the concentration of the agar or gel. Many commercially available growth cabinets and purpose-built walkin growth rooms meet the above specifications. Clearly, an advantage of individual cabinets is that the

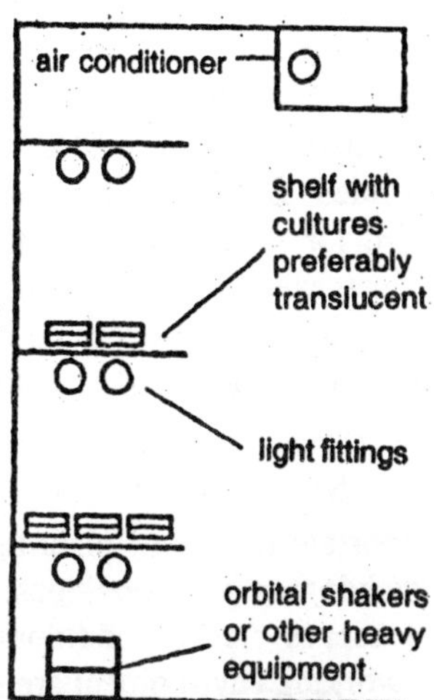

Fig. 5.2: End elevation of a typical plant growth room.

environmental conditions of each can be optimized for growing a particular plant culture. Where suspension cultures are to be grown, an illuminated orbital incubator with temperature control and lights can be used, or a shaking platform with an orbital action can be placed in the culture room.

Culture containers are placed in translucent baskets and arranged on a series of shelves or racks that are illuminated from above by fluorescent tubes with a minimum gap of about 30 cm between lights and the shelf beneath. Where cultures are to be grown in the dark, either light-proof containers can be used or standard containers can be wrapped in aluminium foil. Electrical control gear that is likely to generate heat should be placed outside the room. For reasons of electrical safety, such a culture arrangement is not suitable for growing plants in pots where watering is required.

In addition to the requirements of the culture room, glasshouse facilities may be needed for growing whole plants both before and after tissue culture manipulations. These facilities will also require control of' temperature, lighting and relative humidity. Lighting for whole-plant growth rooms and glasshouses should ideally use horticultural tubes or bulbs to provide the appropriate light spectra.

Where genetically modified material is to be grown, rigorous containment of germplasm must be established and maintained in culture laboratories, culture rooms and glasshouses. It is important that the plant material is not allowed to escape to the environment unless licensed for release. In these circumstances, it must be possible to clean the growth area thoroughly (glasshouse, cabinet or room). Floors and benches should be impervious and crevices where material can accumulate should be avoided. Doors should be secured against casual access and space provided for separate waste management. Openings to the outside should be screened with a mesh fine enough to prevent movement of pollen and aphids. Local licensing regulations should be obeyed in detail in both the design of the facility and the material permitted to be grown.

STERILIZATION

One of the most important factors of plant tissue culture practice is establishing and maintaining aseptic conditions. Therefore, to avoid contamination and to preserve sterile conditions, tools, containers, media, explants and the working environment are sterilized and aseptic techniques are practised.

Sterilizing Transfer Areas and Culture Rooms

Transfer rooms can be sterilized by exposure to UV light. The time of exposure depends on the size of room. Sterilization should only be done when the room is not in use, as UV is harmful to plant material and damaging to eyes. Transfer rooms should also be sterilized by periodic (twice a month) spraying or wiping with commercial brands of bactericides and fungicides. Bench surfaces in transfer rooms should be sterilized before use by wiping with 70% ethanol. Laminar flow hoods are sterilized by wiping the work surface with 70% ethanol and running for 30 min. Culture rooms should be cleaned weekly with detergents and then wiped with 2% solution of sodium hypochlorite or 70% ethanol. Alternatively, floors, walls and work surfaces of culture rooms can be sterilized by treatment with commercial disinfectant such as Lysol.

Sterilizing Instruments and Culture Vessels

Glass bead sterilization

Working pools such as metal instruments are easily sterilized by using a glass bead sterilizer that heats up to 275-350°C. The instruments are inserted into the heated glass beads for 10-60 s to destroy bacterial and fungal spores. The sterilized instruments are then placed in a sterile hood to cool until required.

Dry-heat sterilization

Glassware, metal instruments and materials such as aluminium foil can be sterilized by hot dry air at 130-180°C in a hot air oven for

2-4 h. All items should be sealed or wrapped in foil before sterilization. Please note that paper materials may decompose at higher temperatures.

Autoclaving

Glassware, filters, cotton wool plugs, plastic caps and instruments can be sterilized by autoclaving. Autoclaving is treatment by water vapour at high temperature and high pressure. The standard conditions used for sterilization are 121°C at 1.05 kg cm^{-2} for 15-20 min. Vessels should be closed with a cap or with aluminium foil and instruments should be wrapped in foil or wrapped in paper towels and placed in an autoclave bag. It is important that the steam can penetrate the items for sterilization. Autoclaves in different sizes are commercially available.

Glass pipettes should be plugged with cotton wool and autoclaved or baked in metal pipette cans or glass tubes sealed with cotton wool. Efficiency of autoclaving may be determined using autoclave type. Tips for automatic pipettes should be autoclaved in tip boxes. Instruments and pipettes should be dried in an oven at 80°C before use. Specialized items such as plastic mesh or coated glass slides may be sterilized by soaking overnight in 70% ethanol and drying in a sterile open Petri dish in a laminar flow cabinet for 20 min.

Flame sterilization

Frequently metal instruments previously sterilized by dry heat or autoclaving are removed from their wrapping and re-sterilized just prior to use by dipping in 95% ethanol and placing in the flame of a Bunsen burner. After use, instruments can be re-flamed before reusing. Safety is a major concern with this technique, because of the high flammability of ethanol and the risk of a flash fire. As described earlier, the danger is greater in a flow hood where there is a strong airflow directed towards the worker.

Pre-sterilized containers

A wide range of plastic containers suitable for plant tissue culture is now commercially available. These containers are sold already sterile (sterilized by radiation or ethylene oxide gas) and cannot be re-sterilized. Examples available from Sigma include the Phytacon™, a clear, plastic tub with a clipfit lid and Phytatray™, a clear tray with a raised lid. Some are fitted with filters to allow gas exchange while maintaining sterile conditions (e.g. Magenta™).

Sterilizing Explants

Preparing sterile explants is difficult because the tissue must be treated with disinfectants that are effective in destroying any microbial contamination without harming the explant tissue. The following procedural steps can be used to surface-sterilize explants.

1. Explants are washed in a mild detergent. Leaf and herbaceous tissues may not require this step, but woody tissues and tubers must be washed thoroughly.
2. The washed tissue is thoroughly rinsed with running tap water for 10-30 min.
3. The tissue is then dipped for a few seconds (no longer than a minute) in 70% ethanol.
4. Under sterile conditions the explant is submerged in a disinfectant solution in a sealed bottle and gently agitated for 5-10 min. A wetting agent such as Tween 20 or 80 can be added to the disinfectant to improve contact between disinfectant and tissue. For tissues that are difficult to decontaminate, vacuum infiltration of the disinfectant will enhance the sterilization by removing air pockets.
5. At the end of the sterilization period, the disinfectant is decanted and the explants washed three or more times in sterile distilled water, gently blotted with sterile paper towels and then used to initiate the tissue culture.

Table 5.1. Disinfectants commonly used to sterilize sxplant material for plant tissue culture

Disinfectant	*Concentration (%)*	*Comments*
Sodium hypochlorite	0.5–5	Activity based on available chlorine and oxidizing action
Commercial bleach	10–20	Contains approximately 5% sodium hypochlorite
Calcium hypochlorite	9–10	A solid that must be dissolved and filtered,
Hydrogen peroxide	3–12	Activity based on strong oxidizing action
Benzalkonium chloride	0.01–0.1	Cationic surfactant. Permeates biological membranes
Mercuric chloride	0.1–1.0	Denatures proteins. Very toxic, requires special disposal for waste solutions
Ethanol	70–95	Denatures proteins

For tissues that are difficult to disinfect, it may be necessary to repeat the sterilization procedure after 24-48 h, before selecting the final explant. This allows any remaining microbes time to develop to a stage at which they are more easily destroyed by the disinfectant. Other methods to reduce high contamination of tissue include taking explants from stock plants grown in a greenhouse or growth chambers rather than from plants grown in the field and treating the explant with antibiotics and fungicides. The latter compounds are not always effective and are often phytotoxic.

Sterilizing Culture Media

Two methods are used to sterilize culture media and culture media supplements:

1. Autoclaving for thermostable ingredients;
2. Filter sterilization for thermolabile compounds.

Basal media, ultrapure water and any other heat stable components are placed in glass vessels that are sealed with plastic caps (loosely fitted), aluminium foil or cotton wool plugs and autoclaved at 121°C at 1.05 kg cm^{-2} (100 kPa) for 15-20 min. Large volumes of 2-4 l require a longer time, 30-40 min. Pressures should not exceed 1.05 kg cm^{-2} (100 kPa) as higher pressure can cause the breakdown of carbohydrates in the medium.

Solutions of media components such as vitamins, amino acids and hormones that are thermolabile are filter-sterilized. Sterile filters (e.g. Millipore filters) with a porosity no larger than 0.22 μm are used and the sterile filtrates are collected in sterile containers.

Media Preparation

There is a vast range of media commercially available, which are derived from early experimental work in plant tissue culture. These include the media of Gamborg, Hoagland, Litvay, and Schenk and Hildebrandt. There are also over 20 variants based on Murashige and Skoog basic medium embracing a large number of speciality cultures. Generally, all these media are described as containing the complete range of micro- and macronutrients necessary for plant tissue or cell culture. Others have a reduced component mix to allow the experimenter to vary medium composition as required. Media either can be bought commercially as prepared powder mixtures or can be made up from basic components. In the latter case, great care must be exercised to ensure consistency from batch to batch. If large amounts are being prepared, all solid ingredients need to be finely ground either with a pestle and mortar or with a mill, to ensure homogeneity. Powdered salt mixtures are extremely hygroscopic and should be protected from atmospheric moisture by storing in a desiccator over a drying agent. Alternatively, it may be convenient to prepare inorganic macronutrients and inorganic micronutrients as separate stock solutions.

A major component of some culture media, not normally included in the basal salt mixtures, is the carbon source. This is usually a sugar that can be easily metabolized, like sucrose. Sucrose is normally added at a concentration of 1-3% w/v (10-30 gl^{-1}).

The culture medium may also require single vitamin additions or more commonly, vitamin mixes. These are sold as concentrated, filter sterilized solutions. It is important to choose those that have been previously tested in plant tissue culture: Vitamin mixes such as Gamborg's or Litvay's are also available as powders to be dissolved as 1000× stock solutions. The solutions should be filter-sterilized, aliquoted into 5- or 10-ml sterile tubes and stored at –20°C for subsequent use.

Many plant cultures require growth regulators (hormones). The most frequently used plant hormones are sold as powders to be dissolved. It is useful to prepare a stock solution that is 1000x the final concentration to facilitate small volume additions. Hormones are added before autoclaving or filter-sterilized and added after autoclaving.

It is possible that the medium will require an undefined supplement such as coconut water, casein hydrolysate or banana powder, all of which can be autoclaved. Coconut water, used at 1-10% v/v, is extracted from coconuts. Banana powder is available as lyophilisate and is used at 25-40 gl^{-1}. Casein hydrolysate may be used at 100 mg l^{-1}. Amino acid supplements such as glutamine are very heat-labile and need to be filter-sterilized. Antibiotics or antimycotics should only be added after autoclaving. Again, these are available as plant tissue culture tested and need to be dissolved to a suitable concentration, filter-sterilized and can be stored frozen. Most antibiotics will be stable at –20°C for 3 months; however, Rifampicin should be freshly prepared and Ampicillin is extremely light sensitive.

If solid medium is required, a gelling agent is added before autoclaving. Agar, the most commonly used gelling agent, is available in a range of purities, but for plant tissue cultures, the most pure should be used. Agar is used at a concentration of 0.8 and 2% w/v.

All media should be adjusted to the appropriate pH before autoclaving after all additions (with the exception of agar) have been made. As a general rule, hormones dissolved in alcohol do not affect pH, while other additions are made in such small volumes that the effect is negligible. Prior to autoclaving, flasks containing media should be loosely sealed, for instance with plastic caps, aluminium foil caps, or with cotton or foam bungs.

For safety reasons, a facemask and thick, waterproof gloves should be worn while unloading the autoclave. Although most autoclaves are fitted with safety locks, at the end of the run it is essential to ensure that the temperature is safely below boiling point and that the pressure has returned to atmospheric pressure before opening the autoclave. Failure to do this can result in sudden and spontaneous boiling of the media, which can cause serious burns. It is useful to have a clean area (laminar flow cabinet or similar) in which to allow media to cool after autoclaving and to allow solid media to set. Containers should be level and not disturbed until media have fully cooled. Plates, vessels and liquid media can then be stored for a few days in a refrigerator until required. Solid.media may not be frozen. It is very important that packets of plates and containers are fully labelled with contents, date and experimenter's initials at the time of pouring. It is especially easy to confuse plates of different media prepared at the same time.

Contamination

Cultures should be checked 3-5 days after initiating or subculturing for contamination. Contamination can come from several organisms.

1. Bacteria are the most frequent contaminants. Bacterial contamination is commonly introduced with the explant and is characterized by a slimy appearance. Bacterial contamination can be many colours such as white, cream, pink or yellow.
2. Fungi may enter cultures on tissue explants or from airborne spores. Fungal contaminations are recognized by their 'fuzzy' appearance and occur in many colours.
3. Yeasts are frequent contaminants that may enter cultures on the explant or may be present in the air.
4. Viruses and mycoplasma-like organisms can be sources of contamination and are not easily detected.
5. Insects, such as ants, thrips and mites can be troublesome contaminants in cultures. Thrips emerge from eggs present in the explant and contamination of mites is usually an indicator of poor laboratory hygiene. Thrips and mites are responsible for spreading fungal spores and are often detected through the appearance of fungal growth along the paths of the insects.

Contamination can be introduced in several ways:

- Introduced contamination—resulting from poor laboratory hygiene and/or aseptic technique;
- Initial contamination—due to incomplete sterilization of the explant;
- Latent contamination—usually resulting from endogenous bacteria present in the tissue explant that grow and multiply long after culture initiation. Latent bacteria are a troublesome problem as they may be transferred easily at the time of subculturing.

Positive identification of the contaminating organisms can be established by visual screening or by the process known as indexing. This involves taking a culture sample and inoculating media that are specific for bacteria, fungi or yeasts. The media frequently used are nutrient broth (NB) containing salts, yeast extract and glucose to stimulate bacterial growth, or potato dextrose agar (PDA), for the growth of fungi and yeasts. However, unless cultures are particularly valuable, contaminated material should be disposed of without delay by autoclaving to prevent it spreading to other cultures.

Disposing of Contaminated Waste

Used agar plates should be placed in autoclave bags and the bags placed inside a plate box or metal bucket, designed to withstand the pressures of autoclaving. This arrangement prevents spillage of agar during autoclaving. Used empty Petri dishes, tips, plant material and disposable pipettes may be autoclaved in a bag if local regulations permit. Autoclaved material may be then disposed of as for normal waste. Contaminated liquids may be either autoclaved or treated with freshly prepared sodium hypochlorite (5% final concentration) overnight. These liquids may be disposed of to drains if local regulations permit. Used sharps and needles should be disposed in the correct hazard bins, destined for burning. Used tools such as forceps may be either autoclaved or heated in a bead sterilizer.

Table 5.2. Autoclave conditions for disposing of contaminated waste

Item	*Temperature*	*Time*
Plastic waste (Petri dishes, plate, boxes, no glass)	134°C	10 min
Mixed load plastics and fluids in glass or metal buckets	121°C	35 min
Laboratory coats	134°C	5 min
Fluids in large volumes (5-101)	121°C	At least 30 min

Safety in the Laboratory

Autoclaves operate under high temperatures and pressures, and should be used with great care. Most autoclaves have pressure and temperature locks that will not allow the operator to open the machine until it is completely safe to do so. Even when the autoclave has cooled sufficiently to allow opening the media contained in the bottles may still be very hot and can boil over and spill with the risk of serious burns. For these reasons, insulated, waterproof gloves and a clear full-face visor should be worn when unloading an autoclave. The gloves need to be waterproof because the outside of the bottles or vessels will be very wet and the water will track through a non-waterproofed material, conducting beat through to the operator. The most commonly occurring accidents related to autoclaves are those that involve sealed units, particularly glass bottles. Tightening the lids of screw-cap bottles prior to autoclaving can cause them to explode. Therefore, each bottle must be checked to ensure that the cap is loose enough to allow steam to escape before it is loaded into the autoclave. Autoclaved media in glass bottles should be handled using thick gloves until it has cooled sufficiently to pour. Similar precautions should be taken removing articles from ovens. Another risk in the tissue culture laboratory is that of hazardous chemicals. Care must be taken in the handling and disposal of these according to the COSSH regulations (in the UK) or other appropriate local regulations.

It is important to pay careful attention to the local regulations governing the control of work involving genetic modification (GM). All plant material or equipment that has been in contact with GM material must be autoclaved before it leaves the laboratory. In addition, transformed plant material should only be grown in areas that have pollen and aphid screens on all windows or vents and a double door arrangement to act as a pollen trap. Glasshouse facilities should be divided to separate GM and non-GM plants so that there is no possibility of cross-pollination. Waste materials should be disposed of according to local regulations; living material and waste media should be autoclaved before disposal.

Protocol

Hygiene in the Tissue Culture Laboratory

The tissue culture unit should be designed for easy cleaning. Bench surfaces should be laminated with integrated sinks moulded into them and under-bench cupboards should be raised from the floor.

The floor and skirting should be easy to clean without joints or crevices. Bench-top shelves, up-stands and electrical fittings should be moulded without cracks or gaps. All surfaces should be cleaned regularly with a proprietary disinfectant containing sodium hypochlorite. Where practical, heating and ventilation should cause minimal air-movement, especially near the laminar flow cabinets where conflicting airflow can result in material being contaminated.

Equipment

Laboratory coat reserved for use in tissue culture facility

Laboratory paper towels

Laminar flow cabinet situated in a clean room

Autoclavable waste boxes for Petri dishes

Plastic waste bins lined with autoclave bags for dry plastic waste (e.g. pipettes or tips)

Jugs containing either fresh hypochlorite or other sterilant for contaminated liquids.

Materials and reagents

Spray containing 70% ethanol

Hand wash sink with liquid soap dispenser and elbow taps.

Procedure

1. Leave all personal possessions and coats in a secure area outside the tissue culture laboratory.
2. Tie back hair and wash hands.
3. Put on clean tissue culture lab coat.
4. Thoroughly wipe down laminar flow cabinet with 70% ethanol. Start fan and leave running for 30 min.
5. Meanwhile, spray equipment to be used with alcohol and place in flow cabinet. Ensure that marker pen used to label plates or equipment is not alcohol soluble. If it is, only wipe unmarked parts of plates with tissue soaked in alcohol.
6. The most sterile zone is at the back of the cabinet so use this area.
7. Open media bottles or Peter dishes into the air flow.
8. Discard unwanted equipment or media as soon as convenient. Maintaining sterile conditions is easier when the working area is uncluttered. Autoclave bags for waste should be placed adjacent to the cabinet.

Safety note

Take care when using 70% alcohol spray. Alcohol is toxic and flammable; avoid skin contact. inhalation or ingestion. Keep well away from naked flames or sources of ignition.

Preparing Basal Medium from Commercial Formulations or from Individual Components

The required basal medium can be prepared either by dissolving the commercially available basal salts mix or by using a table of components and preparing the mix from the individual ingredients. An alternative is to prepare macronutrient and micronutrient stock solutions and store these at 4°C until used. Organic compounds should not be stored in solution for more than 2 weeks. Heat-labile compounds such as plant growth regulators, antibiotics or glutamine should be filter-sterilized and added when the medium has been autoclaved and cooled.

Equipment

Balance

Beaker that is at least 50% larger than the volume of medium to be prepared

Spatula

Magnetic stirrer and bar

Measuring cylinder

Glass or plastic bottles for storage (N.B. Glass vessels should not be stored at -20°C)

pH meter

Vessels for autoclaving (conical flasks for liquid media or Duran bottles for solid media)

Autoclave and autoclave tape

Sterile containers to be used for solid media immediately after autoclaving (e.g. Petri dishes or Phytatrays™)

Automatic pipette (1 ml) and tips

Laminar flow cabinet.

Materials and reagents

Ultrapure water

Either a commercially prepared basal salts medium or macronutrient and micronutrient stocks (stored at 4°C) and iron stock.

Organic compounds stock (stored at -20°C)

Sucrose

Gelling agent (e.g. Agar).

1 M NaOH or 1 M HCl to adjust pH.

Procedure

1. Add approximately half the final volume of ultrapure water to the beaker and position it on the magnetic stirrer.
2. Start the magnetic bar spinning.
3. Weigh out the appropriate amount of basal salts or measure the appropriate volume of stock solutions. Add to the water stirring in the beaker.
4. Weigh out the sucrose and add to the beaker.
5. Measure and add any other heat-stable components, such as undefined media supplements.
6. When all solids have dissolved, adjust to final pH, using 0.1 M NaOH or HCl as necessary. Sensitive adjustments may be made using an automatic pipette.
7. Make up to final volume using a measuring cylinder.
8. If preparing liquid medium, measure out into conical flasks (30 ml in a 100 ml flask, or 75-100 ml in a 250 ml flask). The vessels can be capped with aluminium foil sheets, ensuring that no opening is left through which contaminants may enter after autoclaving. Some laboratories also seal flasks with either non-absorbent cotton wool or foam bungs.
9. If preparing solid or semi-solid media, weigh agar (commonly between 8 and 10 gl^{-1}) and add directly to the vessel to be autoclaved. Pour the medium on to the agar, washing the entire agar down to the bottom the flask. Do not over-fill the bottle; for example, 400 ml in a 500 ml bottle is the maximum volume recommended.
10. Solid or semi-solid media can be autoclaved either in the end-use vessel or in a screw-cap bottle. After autoclaving, the media can be cooled and poured into pre-sterilized plastic containers or glass containers sterilized by autoclaving. Few plastics can be autoclaved. If in doubt, check by autoclaving one piece in a glass beaker.
11. Transfer vessels to autoclave. Media vessels sealed with foil can be safely autoclaved but screw-top bottles should be autoclaved with the caps loose. They should be tightened as soon as possible

after the autoclave cycle is complete. Check that everything needed (such as pipette tips, fresh foil caps or forceps) are included.

12. Autoclave at 121 °C (100 kPa, 15 p.s.i.) for 15 min. This may be reduced for vulnerable media such as those rich in sugars. Browning of the medium is a warning sign. The lowest parameters for sterilization are 118°C (70 kPa, 10 p.s.i.) for 10 min. An alternative to autoclaving is to filter-sterilize the most heat-labile component(s) and add them post-autoclaving.
13. After the autoclave cycle has been completed and the chamber has cooled to at least 70°C, the contents may be removed. Tools or tips may need to be dried in an oven prior to use. Media already in end-use vessels should be left, preferably, in a flow cabinet, to cool and (if sold) set. Ensure the vessels are level. If media is to be transferred to a sterile plastic container, it should be allowed to cool until hand hot, when sterile antibiotics or vitamins may be added if necessary, and poured in a laminar flow cabinet.
14. It is important to check that the labels on the media are intact. Petri dishes of media can be stored in the original plastic tubing, labelled with the composition and the date.
15. Media are best stored at 4°C, never frozen and should not be used when more than 2 weeks old.

Note

Hormones and other additions are frequently added to basal media.

Safety note

Tightening the lids of Duran or similar screw-cap bottles prior to autoclaving is dangerous and can lead to them exploding in the autoclave. They should be left resting on the bottle top until the autoclave has been opened and the contents have cooled to less than 70°C. To avoid contamination, the bottles should be transferred to a laminar flow cabinet immediately after they are removed from the autoclave.

Preparation of Hormone Additions to Basal Media

Equipment

Syringe filters (0.22 μM)

10-ml measuring cylinder

Fine balance

Luer-lock syringes (10 ml)

Sterile containers for stock solutions (plastic Universal bottles)

pH meter with fine probe

Pipette

Magnetic stirrer and small stirrer bars.

Procedure

Prepare individual stock solutions for each of the hormones needed in plastic Universal tubes. It is suggested that a 10 mM stock solution is appropriate for most applications. The example of preparing stock solutions of I-naphthaleneacetic acid (NAA) and 6-benzylaminopurine (BAP) will be given.

1. Weigh out 18.62 mg of NAA into a Universal bottle.
2. Add 7 ml of distilled water with stirring (magnetic stirrer).
3. Bring to pH 6.5 with constant stirring by adding 0.5 M KOH dropwise.
4. Transfer to a 10-ml measuring cylinder, bring to volume, mix again.
5. Transfer to a Universal bottle for storage.
6. Weigh out 22.52 mg of BAP into a second Universal bottle.
7. Add 7 ml of 10% aqueous ethanol with stirring (magnetic stirrer).
8. When fully dissolved, transfer to a 10-ml volumetric flask and bring to volume with ultrapure water.
9. Filter both solutions through 0.22 μm syringe filters into sterile plastic Universal bottles under sterile conditions in a laminar flow cabinet.
10. Use immediately or store at –20°C.
11. To use after freezing, thaw and take appropriate aliquot aseptically before adding to medium.
12. Loosen the tops of both the sterile growth regulator bottle and the container of freshly autoclaved medium.
13. Push the automatic pipettor into the top of a sterile tip presented in a tip box and lift the tip out without touching the sides.
14. Using the free hand, lift the tip off the bottle containing the additive and withdraw the set amount.
15. Replace the lid loosely on the bottle and use the same hand to remove the top of the bottle of medium.
16. Dispense the additive into the medium being very careful not to touch anything.
17. Discard the tip into a waste container.

18. Cap the medium bottle and mix thoroughly by rotating horizontally on the bench. It is not the best sterile practise to invert the bottle.
19. Liquid medium can be stored or used while solid medium can be poured and allowed to set, Preparation of coconut water supplement for basal media.

Preparation of Coconut Water Supplement for Basal Media

Equipment

Hand or electric drill with 8-mm wood bit Bench vice

Funnel (500-ml)

Filter paper (Whatman number 1)

Retort stand, clamps and bosses

Conical flasks

Plastic bottles for storage (100-200-ml).

Procedure

1. Purchase a sack of good-quality fresh ripe coconuts from a wholesaler.
2. Mount a coconut in a bench vice and drill two 8-mm holes through the shell.
3. Drain the coconut water (liquid endosperm) via the filter funnel and Whatman number 1 filter paper into a conical flask.
4. Check the product of each coconut by eye and smell for quality. A yellow colour or an acid smell is a contraindicator.
5. Pool the coconut water from the entire batch and mix.
6. Store frozen at –20°C.

6

Callus Cultures

Callus is an amorphous mass of unorganized thin-walled parenchyma cells. When a plant is wounded, callus formation occurs at the cut surfaces and is thought to be a protective response by the plant to seal off damaged tissues. The formation of wound callus has been observed in almost all groups of living plants.

In culture, callus is initiated by placing a fragment of plant tissue (an explant) on solid culture media under aseptic conditions. Callus is induced and formed from proliferating cells at the cut surface of the explant tissue. Depending on the species, callus can be initiated from a variety of tissues by employing the appropriate growth medium. However, rapid cell division can be more easily induced in some tissue than in others. The *in vitro* formation and proliferation of callus is enhanced by the presence in the medium of hormones (auxins and cytokinins) that promotes cell division and elongation.

Origin of Callus

During the *in vitro* initiation of callus, the cell differentiation and specialization that occurred in the parent plant is reversed and cells of the explant become dedifferentiated. The process of dedifferentiation is characterized by changes in metabolic activity, the disappearance of storage products and rapid cell division that gives rise to undifferentiated and unorganized parenchyma cells. The lack of structural organization persists as the callus grows, although a homogeneous callus consisting entirely of parenchyma cells is rarely observed. As growth of the callus proceeds, centres of meristematic activity are formed. Random and rudimentary cambial zones may give rise to regions showing vascular differentiation in the form of sieve elements, suberized cells or tracheary elements. Some of the centres of meristematic activity form nodules that may be precursors of shoot apices, root primordia or incipient embryos, which are capable of further development if exposed to suitable culture media.

Types of Callus

Callus varies widely in its general appearance and in other physical features. The variation depends on the parent tissue, the age of the callus and the growth conditions. Callus may be white, green or highly coloured due to the presence of anthocyanin pigments. Callus may consist of loosely packed cells and be friable (i.e. easily crumbled or fragmented), or may be lignified, with densely packed cells and hard in texture (non-friable). Furthermore, callus may or may not be embryogenic, that is, able to form embryos either spontaneously or when grown under suitable conditions. In monocotyledonous grasses like maize, these features have been used to define two types of callus. Type I callus is non-friable, regenerates somatic embryos and organs and frequently produces leaf-like structures. Type II callus is friable. undifferentiated and regenerates only somatic embryos.

Callus cultures are also predisposed to genetic instability and as a result, variation in phenotype within the same culture may occur. This phenomenon known as *somaclonal variation.*

Role of Callus in Embryogenesis, Organogenesis and Cell Culture

Formation of callus is a fundamental step in the *in vitro* culture of many types of plant cells and tissues and in some methods of plant genetic manipulation. For example, *in vitro* callus provides the most frequently used totipotent cells from which whole plants are regenerated via either organogenesis or somatic embryogenesis. Callus is often used as the target tissue for genetic transformation and callus formation is initiated for the regeneration of plants after transformation of other tissues. Dispersal of friable callus into single cells or clumps of cells is used universally as the method of initiating cell suspension cultures.

Initiation and Establishment of Callus Cultures

The successful establishment of a callus culture requires consideration of four important areas:

1. Selection of suitable parent material;
2. Choice of explant and method of isolation;
3. The media and culture conditions required;
4. Optimization of the culture conditions.

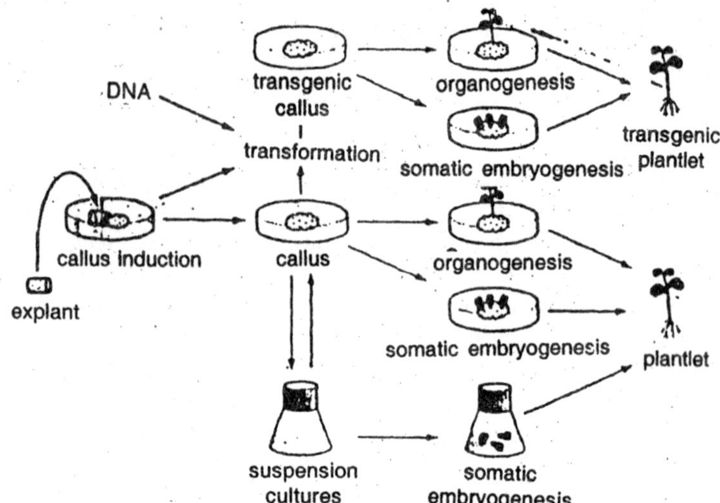

Fig. 6.1. Alternative applications of callus in plant tissue culture and genetic transformation.

Selection of suitable starting material requires seedlings or plants that are growing vigorously and are free of disease. If the callus culture is to be used as part of a micropropagation programme, the parent plant must show the desired genetic characteristics. Such a plant is known as an elite mother plant.

Explants are isolated from elite material, in sterile conditions. The mother plant material and explants are decontaminated and surface sterilized. Explants are removed with sterile instruments. The method for removing the explant is unimportant, though the act of wounding is important in producing a wound response. Leaf pieces can be cut using sterile scissors, a scalpel or a cork-borer. Stem and storage root cores are usually taken with a cork borer and then cut transversely with a sterile scalpel. Explants taken from internal plant tissues, for instance, cores taken from within the stem (pith) or from taproots, may only require brief surface sterilization after removal from the plant. In each case, care must be taken to keep the explant sterile throughout and to discard any plant material that may be contaminated.

The formation of callus is usually achieved by placing the sterile explant onto an appropriate solid growth medium in a container such as a Petri dish or culture tube. The container is closed and incubated under the prescribed environmental conditions. After a period of 2-3 weeks, callus formation is observed first as outgrowths at the edges of the wounded material, and later as a mass of cells which gradually grows over the original plant material. Callus formation may be preceded by expansion growth of the plant tissue.

Table 6.1. Growth 'square' to test media concentrations of auxin and cytokinin for optimal callus growth

	Kinetin (μM)				
IAA (μM)	*0*	*0.5*	*2.5*	*5*	*10*
0	1	2	3	4	5
0.5	6	7	8	9	10
2.5	11	12	13	14	15
5	16	17	18	19	20
10	21	22	23	24	25

Optimization of the growth of the callus may require initial trials with several different media; transfer of established callus to new media with varied compositions, or selection of light-coloured (no browning) friable fragments for subculture. Examples of media for the optimal growth of callus from several species are given in the protocols of this chapter. If callus production from other species is required, it is advisable first to undertake a literature search to see whether media have previously been optimized for that species. If no information can be found, use one of the standard media and set up a number of agar plates with varying auxin and cytokinin concentrations. One of the best ways to do this is to set up a growth (Latin) 'square' of 25 plates, with five auxin concentrations in combination with five cytokinin concentrations. For this growth trial, the inocula of callus should be of uniform size and large enough (about 20 mg) to ensure growth on a new medium.

Protocols are presented in this chapter for callus production from *Arabidopsis thaliana* seedling from imbibed seeds of gymnosperms and angiosperms and from roots and shoots of monocots and dicots, using methods practised in our laboratory.

Monitoring the Growth of Callus

The growth of callus is monitored by measuring the fresh weight and dry weight of the cultures. For fresh weight measurements, the callus is transferred to a pre-weighed foil weighing boat and rapidly weighed. It is important that weighing of the callus is done as quickly as possible after removal from the container as the culture loses water rapidly in open air. After the fresh weight measurement, the callus and boat is then oven dried (60°C) overnight and re-weighed to obtain a dry weight. As these measurements result in the death of the culture, they are therefore only useful for optimizing growth conditions. Rough estimates of growth may be made by estimating the diameter of the callus *in situ* and the health of the culture can be assessed by observing its colour. When the medium is unsuitable, growth is slow and browning of the culture occurs. Usually, rapid growth and a light coloured callus indicate a healthy culture. A friable callus of crumbly appearance is very suitable for breaking up, either for subculturing or to produce a suspension culture. In our experience, darkening of the callus is often associated with cell death and it is therefore best to avoid such material. Sometimes discolouration of the culture is an indication of microbial contamination, in which case the culture should be disposed of by autoclaving.

Callus cultures are normally maintained at around 22-25°C under lowintensity fluorescent light with a dark/light cycle of 8 h:16 h. Callus cultures can be maintained under these conditions for several years with subculturing every 3-6 weeks depending on the species and the growth rate of the culture. Callus cultures are often used as a means of preserving germplasm by storage and maintenance under slow-growth conditions.

Genetic Transformation of Callus

Callus cultures are important elements of alma tall current practical transformation strategies. Callus cells are frequently used as targets for transformation both by microprojectile bombardment and by *Agrobacterium tumefaciens*. In addition, when other material such as leaves or embryos are used as targets for transformation, callus is initiated either before transformation or during the recovery of transformed cells. After transformation, a callus stage usually precedes the differentiation of roots, shoots or embryos during the induction of plant regeneration.

Agrobacterium-mediated transformation of dicots utilizes sterile explants of plant material such as leaves, cotyledons, stems, germinating seeds and callus cultures. Typically, leaf discs or pieces are used as the material for transformation and may be transformed without or with prior induction of callus. Leaf pieces are incubated with the bacteria for 5-10 min to allow the bacteria to infiltrate the tissue, then blotted dry and transferred to a fresh plate of solid medium where the leaf tissue and bacteria are co-cultivated for 2 days. By the end of this period, the bacteria are seen to overgrow the leaf surface. The leaf discs are washed clear of bacteria and then transferred to a medium containing antibiotics (e.g. carbenicillin and timentin) to kill the *Agrobacterium* and antibiotics (e.g. hygromycin) to select transformed cells. Discs are transferred at weekly intervals to new media and by adjusting the hormonal composition of the media, transgenic callus and/or plantlets can be induced to form.

Monocot callus (e.g. maize) derived from embryogenic tissue is the commonly used material for direct transformation by microprojectile bombardment.

PROTOCOL

Preparation of Plant Material and Explants

Equipment

Sterile:

- Conical flask (250 ml), sealed with aluminium foil
- Petri dish
- Tea strainer or muslin squares, autoclaved in foil packets
- Sterile containers for sterilizing plant material (e.g. beaker sealed with foil cap)
- Sterile tips for automatic pipettor
- Universal bottles or suitable sterile germination vessel Spatula
- Forceps
- Scalpel
- Cork borer
- Small dissecting scissors.

Non-sterile:

- Binocular dissecting microscope
- Automatic pipettor.

Materials and reagents

- Commercial bleach diluted to contain 6% available chlorine; this can also be made up from an aqueous laboratory preparation of sodium hypochlorite. Do not autoclave
- Absolute alcohol
- Surfactant (e.g.Tween 80)
- Sterile distilled or deionized water
- Germination medium (e.g. MS medium with 2% sucrose, 1.0 mg l^{-1} thiamine, 0.5 mg l^{-1} pyridoxine, 0.5 mg l^{-1} nicotinic acid,

0.5 g l^{-1} 2-(*N*-morpholino) ethanesulphonic acid (MES) pH 5.7 with 1 M KOH, 0.8% Bacto agar). Sterilized by autoclaving

Source of plant material (e.g. leaf or seed).

Procedure A—Germinating seedlings for callus production

1. Place the seeds in the tea strainer, make a small muslin bag to hold them or, if large, place them direct into a 250-ml conical flask.
2. Submerge them in absolute alcohol for a few seconds. Pipette off alcohol and discard.
3. Immerse the seeds in hypochlorite with a few drops of surfactant for 15-20 min (the duration of immersion can be varied—less for more sensitive seeds, longer if infection is a problem).
4. Rinse the seeds 3 × 5 min with 250 ml of sterile distilled water. If seeds float to surface, either pipette off liquid or try a low-speed centrifugation step (e.g. 1000 g for 5 min).
5. Working in aseptic conditions, place the seeds in a sterile Petri dish, then transfer them to germination medium. In either a Universal bottle or other suitable sterile germination vessel. For *Arabidopsis*, pH appears to be very important; we always germinate it in a medium of less than pH 6. Seeds may be transferred by using an automatic pipette with the end of the tip removed.
6. The seedlings can then be germinated in a suitable incubator (temperature and lighting will depend on species; both species requiring a 16 h light : 8 h dark cycle and 22°C for *Arabidopsis* and 25°C for *Nicotiano*) and used to provide sterile material from which callus can be regenerated.

Procedure B—Explants from plant material for callus production

This method will work with a range of plant parts from plants grown in a growth chamber, glasshouse or in the field. Stem leaves and other parts may all be regenerated. Root material presents more of a problem unless grown under aseptic conditions (see above) though cores bored from tubers or large roots using a sterile cork borer can produce successful callus.

1. Wash the plant material in water and dip it briefly in absolute ethanol.
2. Immerse It In hypochlorite with a few drops of surfactant added for 5-10 min (the duration of immersion can be varied—less for more sensitive material, longer if infection is a problem).
3. From now on, work in a laminar flow cabinet.

4. If necessary, use a sterile scalpel to cut away any surface material and repeat step 2; for instance. removing outer leaves or trimming away cut ends will all help to remove contaminants. Work on a sterile surface, like the inside of a Petri dish lid.
5. Rinse the tissue 3 × 5 min in sterile distilled water.
6. Now take the tissue and cut away the outer layers that have been In contact with the hypochlorite. Work on a sterile surface, using freshly sterilized instruments.
7. Cut the tissue into small sections up to 10 mm diameter and 2-3 mm thickness and place them onto callus medium.

Procedure C—From embryos from imbibed seeds

This protocol is described for a gymnosperm, Norway spruce; however, it can be used for any seed large enough to be dissected.

1. Imbibe the seeds in water overnight. While mature seeds can be used, immature seeds may yield more viable cultures.
2. Surface sterilize the seeds, either in 5% sodium hypochlorite for 10 min, or 15% hydrogen peroxide for 15-20 min.
3. Rinse the seeds 3 × 250 ml of sterile distilled water.
4. Using aseptic technique and a binocular microscope carefully dissect the seeds and excise the embryo by squeezing the sides of the seed with fine forceps. The embryo will slip out of the endosperm and can be picked up with the fine forceps.

Embryos are ready for callus culture without further sterilization; between 10 and 20 are required per agar plate.

Notes

A very similar method can be used for maize callus; however, it is important to obtain embryos 12.5-17.5 mm long, 9-11 days after pollination. The entire ear is sterilized by immersion in sodium hypochlorite as above. The kernel is cut from the silk scar to the base and the embryo removed and placed onto callus induction medium.

Callus from a Dicot Root: *Arabidopsis thaliana* and *Nicotiana tabacum*

Equipment

Sterile:

Petri dishes

Spatula

Scalpel.

Non-sterile:

Growth cabinet (22°C, continuous white light or 16 h light, 8 h dark)

Parafilm™ or 3M Micropore gas-permeable tape (this allows gas exchange and enhances growth).

Materials and reagents

Callus induction medium (sterile, in 6 cm Petri dishes)

Gamborg's B5 basal medium (Sigma G5768) with 2% glucose, 0.8%Agar, 0.5 gl^{-1} MES, 0.5 mg l^{-1} 2,4-D, 0.05 mg l^{-1} kinetin, pH 5.7.

Plant material—seedlings germinated in aseptic conditions.

Procedure

(Work in a laminar flow hood):

1. Transfer seedlings to a sterile surface (e.g. the inside of a Petri dish lid).
2. Excise the roots and section them into roughly 1 mm lengths
3. Carefully transfer the cut pieces to the surface of the callus induction medium using sterile spatula.
4. Seal the Petri dishes around the edge with a piece of Parafilm™ or gaspermeable tape and place in an incubator at 22°C with continuous white light or 16 h light, 8 h dark.
5. Monitor over 2-3 weeks. Growth of callus will reach 3-5 mm over this period.
6. Callus may be subcultured by removing it and placing onto fresh medium periodically every 4-6 weeks). It may be subdivided using a sterile scalpel to generate more individual calli.

Callus from Dicot Shoot/Leaf—*Arabidopsis thaliana* and *Nicotiana tobacum*

Equipment

Sterile:

Petri dishes

Spatula

Scalpel.

Non-sterile:

Parafilm™ or 3M Micropore gas-permeable tape

Growth cabinet (22°C or 25°C depending on species, continuous white light or 16 h light : 8 h dark cycle).

Materials and reagents

Arabidopsis callus induction medium (sterile, in 6-cm Petri dishes)

Nicotiana callus induction medium (sterile, in 6-cm Petri dishes)

MS medium with 3% sucrose, 0.8% agar, 1 mg l^{-1} 2,4-D, 1 mg l^{-1} kinetin, pH 5.2.

Plant material—surface sterilized leaves or seedlings germinated in aseptic.

Procedure

(Work in a laminar flow hood)

1. Excise 5-10 mm diameter segments of leaf material from expanding leaves.
2. Carefully transfer the cut pieces to the surface of the callus induction medium using a sterile spatula (it does not matter which way up the leaf pieces are placed).
3. Seal the Petri dishes around the edge with Parafilm™ or gas-permeable tape and place in an incubator at 25°C (for *N. tabacum*) and 22°C (for *Arabidopsis*) with continuous white light or a 16 h light : 8 h dark cycle.
4. Monitor over 2-3 weeks. Callus will begin to form at the edge of the cut leaf, it will be ready for reculturing on the same medium after 6 weeks and can be maintained by subculturing every 4 weeks. Callus may be subdivided using a sterile scalpel to generate more individual calli.

Callus from a Monocot (e.g. Maize, Rice)

While successful callus may be generated from roots and shoots, callus derived from immature embryos has the best potential for embryogenesis and the production of suspension cultures for protoplasts for transformations. The protocol is given with variations for rice and maize.

Equipment

Sterile:

Petri dishes

Spatula.

Non-sterile:

Parafilm™ or 3M Micropore gas-permeable tape

Growth cabinet (25°C, dark, or seal Petri dishes in aluminium foil).

Materials

Induction medium (sterile, in 6-cm Petri dishes): 4 g l^{-1} salts. 1 ml l^{-1} Eriksson's Vitamin mix, 0.5 mg l^{-1} thiamine-HCl, 0.1 g l^{-1} vitamin-free casamino acids. Supplement with 2 mg l^{-1} 2,4-D, 20 g l^{-1}, sucrose, 0.8% agar, pH 5.3 (maize) or 4.4 mg l^{-1} 2,4-D, 50 g l^{-1} sucrose and 1.0% agar, pH 5.8 (rice).

Plant material—immature maize embryos from kernels, use embryos from dehulled rehydrated rice seeds.

Procedure

(Work in a laminar flow hood):

1. Separate the embryo, from the endosperm and place onto callus initiation medium, with the embryo axis in contact with the medium, scutellar side up. Ten embryos may be placed per plate.
2. Seal plate edges with Parafilm™ or 3M gas-permeable tape and incubate In the dark at 25°C.
3. Maize: observe over 2-3 weeks; when a friable, soft embryogenic callus has formed, transfer it to the plates containing the same medium as previously. The callus formed has numerous small embryoids protruding from its surface. Rice: observe growth and transfer to fresh media every 10-14 days.
4. If embryoids begin to develop further, increase the 2,4-D concentration or decrease the time between subculturing to maintain the culture as callus.

Callus from Gymnosperms

Equipment

Sterile:

- Petri dishes
- Spatula

Non-sterile:

- Parafilm™ or 3M Micropore gas-permeable tape
- Growth cabinet (20-25°C, dark, or seal Petri dishes in aluminium foil).

Materials and reagents

Norway spruce callus induction medium

0.5 strength Litvay's medium with 1% sucrose, 2 mg l^{-1} 2,4-D, 1 mg l^{-1} benzyladenine, 500 mg l^{-1} casein hydrolysate, 250 mg l^{-1} glutamine, pH 5.6.0.8% agar.

Plant material—embryos from imbibed seeds.

Procedure

(Work in a laminar flow hood):

1. Place 10-15 embryos onto callus induction medium using a sterile spatula.
2. Seal the Petri dishes around the edge with a piece of Parafilm™ or gas-permeable tape and place in an incubator at 20-25°C in the dark.
3. Transfer to fresh medium after 2-4 weeks.

   Monitor weekly; the. culture can be maintained by subculturing every 4 weeks. Callus may be subdivided using a sterile scalpel to generate more individual calli that are placed onto new agar plates.

Transforming Maize Callus by Particle Bombardment

Equipment

Sterile:

Petri dishes

Spatula

Scalpel.

Non-sterile:

Parafilm™ or 3M Micropore gas-permeable tape

Growth cabinet (22°C, continuous white light or 16 h light : 8 h dark cycle)

Particle gun (e.g. Biorad).

Materials and reagents

These media must all be sterile.

Chu N6 agarose medium

Selection media: agarose plates as above supplemented with selection antibiotic or herbicide for the selectable marker.

Microprojectile/DNA preparation, including a selectable marker in the expression cassette.

Maize callus, less than 1 week old.

Procedure

(Work in a laminar flow hood):

1. Prepare a circle of embryogenic callus about 5 cm in diameter on Chu N6 agarose in a 25-cm Petri dish. Try to make the layer as thin as possible as the particles will only penetrate the upper cell layers adequately.

2. Place the Petri dish under the particle gun, align it with the callus and bombard it with microparticles according to the manufacturer's instructions. Optimization may be usefully carried out using a transient expression system.
3. Place the callus onto a selection medium and leave for 1-2 weeks.
4. Transfer the callus to fresh plates with selective media. Grow for 1-2 weeks further. One initial plate should be spread thinly onto a number of new plates.
5. Some calli will grow better than the others in the presence of the selection medium. These calli are transgenic, and contain both the selectable marker gene and the gene of interest. Place them onto fresh selective plates and grow for a further 4 weeks. Confirm the presence of the transgene at this stage by molecular techniques.
6. At this stage, after 10-12 weeks, the calli are ready to be plated out for the regeneration of plants or for fragmenting to form transgenic suspension cultures.

Preparation of a Transgenic *Arabidopsis thaliana* Callus using *Agrobacterium*

Equipment

Sterile:

Petri dishes

Spatula Forceps

Scalpel.

Non-sterile:

Growth cabinet (22°C, continuous white light or 16 h light : 8 h dark) Parafilm™ or 3M Micropore gas-permeable tape.

Materials and reagents

All media need to be sterile

1. Half strength MS liquid medium, 2% sucrose (to calculate, will need 25 ml per Petri dish. If using 10 plates prepare 250-300 ml).
2. Ten plates without antibiotics—half strength MS medium, 2% sucrose, 1% agar, 0.8 mg l^{-1} BAP, 0.1 mg l^{-1} IBA.
3. Ten plates as 2 above with antibiotics; 100 μg ml^{-1} carbenicillin, 20 μg ml^{-1} timentin (ticarcillin/clavulanic acid) and selective antibiotics (e.g. 40 μg ml^{-1} hygromycin).
4. Ten plates of callus induction medium.

Procedure

(Work in a laminar flow hood):

1. Prepare callus leaf pieces.
2. Float the pieces on 25 ml of MS medium + 2% sucrose + 400 μl 24 h *Agrobacterium* culture in a sterile Petri dish.
3. Gently agitate for 20 min at 28°C and 50 rpm in an orbital incubator in the dark.
4. Take squares out of liquid medium and place onto antibiotic-free plates leaving a 1 mm gap between leaf pieces.
5. Wrap plates in foil.
6. Incubate for 3 days at 22°C in the dark.
7. Move the leaf squares to antibiotic plates and seal edges with Parafilm™ or 3M Micropore gas-permeable tape.
8. Transfer leaf squares to fresh antibiotic plates at weekly intervals. If shoot formation is desired, continue to use the plates described under Materials and Reagents item 3 above. If callus production is desired, use the plates described in Materials and Reagents, item 4 above.
9. Subculture the calli at regular intervals to fresh plates.

7

PROTOPLAST CULTURE

The plant cell wall is a multi-layered structure composed of polysaccharides and proteins. The polysaccharides are cellulose, a polymer of glucose; pectic compounds, which are polymers of galacturonic acid molecules; and hemicellulose a polymer consisting of a variety of sugars including xylose, arabinose and mannose. Cell walls consist of three types of layers.

1. The middle lamella—the first layer formed during cell division, forming the outer surface of the wall and is shared by adjacent cells. It is composed of pectic compounds and proteins.
2. The primary wall—is formed after the middle lamella consisting of a framework of cellulose microfibrils embedded in a gel-like matrix of pectic compounds, *hemicellulose* and *glycoproteins*.
3. The secondary wall—formed after cell enlargement, a rigid structure composed of cellulose, hemicellulose and lignin.

Cell walls fulfil several important roles, but two key functions are: (i) to provide tensile strength and limited plasticity so that the cell can develop high turgor pressures without rupturing. Turgor pressure provides support for non-woody plants. (ii) To provide a tough physical barrier that protects the interior of the plant cell from invading micro-organisms. The pore sizes of the wall are small enough to exclude even the passage of viruses. Microbes that are saprophytes and pathogens of plants have a range of hydrolytic enzymes that they use to degrade the cell wall and gain entry to the cell's interior.

The properties of the plant cell wall that make it an efficient protective barrier *ipso facto* restrict the use of plant cells for a variety of cell and tissue culture techniques, including the delivery of large molecules into the cell, somatic hybridization and genetic manipulation. These techniques can therefore only be applied to plant cells after removal of the cell wall.

Careful removal of the plant cell wall results in a viable, spherical cell called a *protoplast*. Protoplasts consist of the original cell's contents bounded by the plasma membrane. With appropriate culture conditions protoplasts will resynthesize the cell wall and undergo cell division to form callus from which new plants can be regenerated.

Isolation of Protoplasts

Protoplasts can be isolated from a variety of whole-plant tissues and from plant tissue cultures like callus and suspension cultures. Removal of the cell wall for protoplast isolation is achieved either by mechanical means or by enzymatic digestion.

Mechanical Isolation

The essence of the mechanical technique is to subject plasmolysed tissue to a number of sharp cuts followed by de-plasmolysis to release the protoplast from the cut ends of cells. This technique, which was one of the earliest methods attempted, is difficult to apply and the

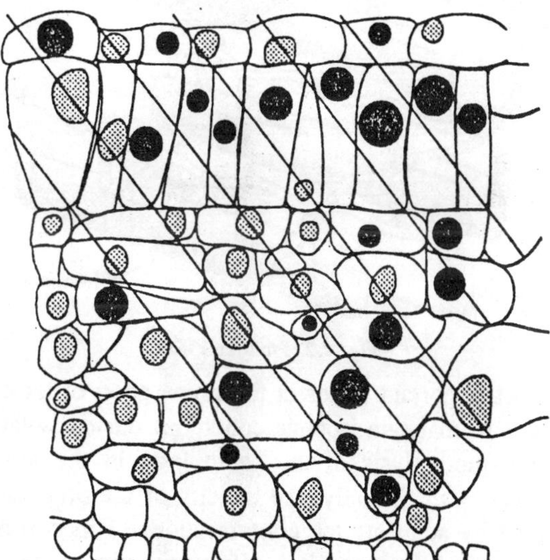

Fig. 7.1. Mechanical isolation of protoplasts. Diagram of part of a transverse section of a leaf showing plasmolysed cells containing protoplasts. The diagonal lines represent scalpel cuts which enable intact protoplasts (black) to be liberated, provided that a corner of the cell wall has been cut off. Speckled protoplasts are either still entrapped or damaged.

yield of protoplasts is meagre. For practical reasons, this method is rarely used except for special situations where wall-degrading enzymes have an unavoidable deleterious effect on the protoplasts and only small quantities of protoplasts are required.

Isolation by Enzymatic Digestion of the Cell Wall

Isolation of protoplasts by the enzymatic digestion of the cell wall was first introduced by Cocking in 1960 and is now the universally adopted method. A combination of cellulases, hemicellulases and pectinases are used to break down the cell wall. The enzymes are obtained from culture filtrates of microorganisms that degrade cell wall material. A commonly used cellulase (Onozuka R10) comes from the filamentous fungus *Trichoderma reesei* and contains an assortment cellulases and hemicellulases. The most frequently used pectinase is Macerozyme and alternatives are Pectolyase and Pectinase.

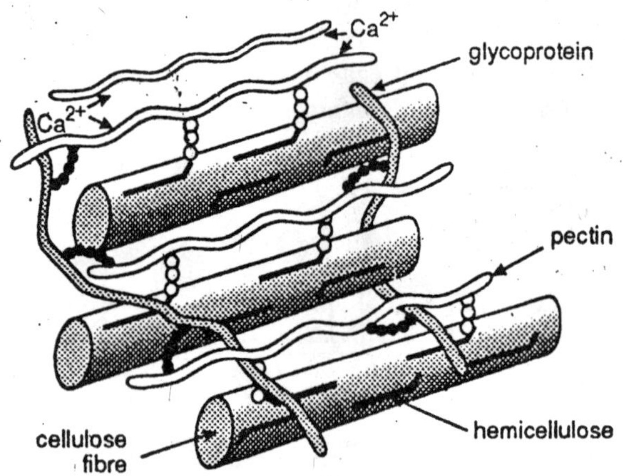

Fig. 7.2. Plant cell walls.

The most important factor in the preparation of intact and viable protoplasts is to maintain isotonic conditions during isolation in order to prevent osmotic lysis of the protoplasts in the absence of the supporting cell wall. Usually, the tonicity of the preparation medium is maintained by adjusting the concentrations of inert osmostabilizers such as mannitol and sorbitol. However, in some procedures, metabolizable sugars such as sucrose are also added.

The general procedure involves incubating the ant tissue in a salt solution containing the hydrolytic enzymes and a suitable osmostabilizer. The digestion medium contains calcium ions to aid-membrane stability and a zwitterionic buffer to control pH, especially pH changes that may occur as a result of cell lysis. Generally, the concentration of the mannitol or sorbitol is such that incipient plasmolysis (shrinking of the protoplasts from the cell wall) occurs. It is not uncommon to expose the source tissue to a preliminary plasmolysis incubation step before addition of the wall-degrading enzymes. However, this step is not included in some recent procedures, particularly those requiring long enzymatic digestions.

When the source tissue is stem or root, the material is cut into sections. Leaves are cut into strips, or for leaf mesophyll protoplasts, the epidermis is removed with the aid of tweezers and the stripped leaves are then exposed to the incubation medium. Penetration of the digestion medium into the tissues is sometimes helped by infiltration under vacuum. If the source is callus the material is broken into small pieces. Suspension cultured cells are harvested, washed and resuspended in the appropriate incubation medium. The length of time the tissue is exposed to the hydrolytic enzymes depends on the tissue and on the concentration of enzymes used. With lower concentration of enzymes, incubations proceed for 12-14 h, whereas with higher concentrations of enzymes the total incubation time is usually 2-3 h.

Purification of Isolated Protoplasts

At the end of the digestion period, the isolated protoplasts have to be separated from the enzymes and cellular debris and transferred to a suitable medium. This purification step is accomplished by gentle centrifugation, filtration through a nylon mesh (approximately 60-70 μm) followed by centrifugation or by the use of a density gradient centrifugation step.

Protoplast Viability and Density

The viability of the isolated protoplasts can be determined by the same viability tests as used for cells in suspension culture. Exclusion of Evans blue dye detected by microscopy or accumulation of FDA visualized by fluorescent microscopy are standard viability tests.

The density of protoplasts in the suspension (number/unit volume) is usually determined by counting with a modified haemocytometer such as an improved Neubauer or Fuchs-Rosenthal with a field depth of 0.2 mm.

Protoplast Culture

Isolated protoplasts can be cultured in an appropriate medium to reform cell walls and generate callus. In determining the optimal culture conditions, it is important: (i) to determine the optimal density for culture. If the density is too low, the protoplasts will lose soluble cell components because of the relatively high volume of surrounding medium and fail to divide. (ii) To establish the optimal auxin to cytokinin ratio. This can be determined by the growth square arrangement. (iii) To maintain the osmoprotectant in the medium until the cell wall has reformed. The concentration can be reduced incrementally with successive subcultures as the cells grow and divide.

Protoplasts can be cultured in the following ways.

1. Hanging-drop cultures. This involves culturing protoplasts in droplets (100 μl) suspended on the lid of a Petri dish, with sterile water in the base of the dish to provide humidity. Because of the small volumes required this arrangement is a convenient way of establishing optimal conditions of growth.
2. In the wells of microtitre plates.
3. On a semi-solid medium using agarose plates. This is an effective method as it provides a supporting matrix for the protoplasts. Standard agar is toxic to protoplasts and low-temperature gelling agarose is used instead. The protoplasts are either layered on the surface of the agarose or embedded in the agarose by suspending in a small volume of the gel at 40°C and added to a small Petri dish before it sets. Alternatively, the protoplasts-gel suspension is layered on a Millipore filter or a nylon membrane to set. The protoplasts embedded in the gel on the membrane are cultured over a nurse culture. The nurse culture normally consists of actively dividing non-embryogenic cultured cells embedded in agarose gel. The nurse culture is thought to provide the protoplasts with growth factors that stimulate wall synthesis and cell division.

Protoplasts from some species begin cell wall regeneration within a few hours of isolation whereas others may take several days to complete the new wall. The formation of the wall can be followed by using the compound Calcofluor White. This dye fluoresces when it binds to wall material and can be detected by fluorescent microscopy. Once the protoplasts have formed a new wall, they undergo cell division and callus is generated.

Uses of Protoplasts

Protoplasts are a single cell system that is useful for basic research as well as applied plant science. Protoplasts can be used in plant cell metabolic studies including photosynthesis. They are ideal for studies on cell wall synthesis and deposition and are the starting point for the isolation of organelles like vacuoles and nuclei. In addition, protoplasts are the most amenable higher plant system for the application of flow cytometry. Flow cytometry is a powerful technique that uses specific fluorescent labelling of cellular components and has found wide application in studies on mammalian cells. By using endogenous fluorescent molecules or added fluorescent tags, flow cytometry can be used for cellular analysis and cell sorting of plant protoplasts.

Protoplasts can be transformed by *Agrobacterium*, and without the cell wall barrier they are a convenient cell system for genetic transformation by direct methods of DNA transfer. Direct transfer of plasmid DNA into protoplasts is achieved by:

1. Electroporation—transient pores are induced in the plasma membrane of protoplasts by short, high-voltage pulses. Some of the pores are large enough to allow the uptake of DNA and will reseal spontaneously trapping the DNA in the protoplast;

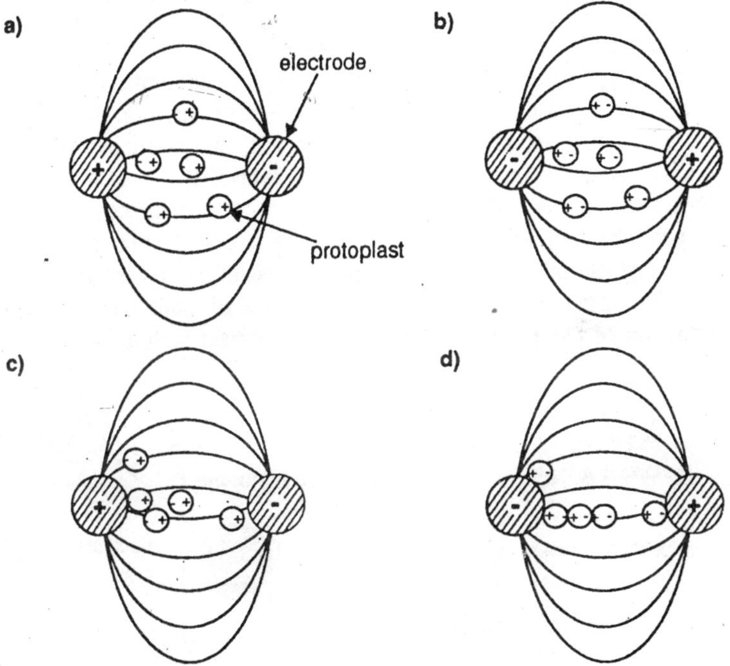

Fig. 7.3. Forces in electrophoresis. (a) An AC current is applied across the electrodes. This induces the protoplast to become dipoler; (b) half cycle later, the electric potential is reversed and dipoles on the protoplasts are also reversed. The cycle is repeated (c and d) many times. The dipoles on the protoplasts attract adjacent protoplasts to each other (c) such that protoplasts begin to form short 'pearl chains' (d).

2. Chemical procedures—based on the addition of calcium or PEG. The use of PEG methods also provides a useful way of directing DNA uptake into chloroplasts.

Protoplast transformation by direct DNA transfer is a very effective strategy to obtain transient gene expression. The technique of transient expression is increasingly being used in basic research to study a range of plant biological problems, because it does not suffer the disadvantages of delay and environmental risks that are associated with stable transformation systems. However, in transformation work aimed at improving cultivars, techniques involving protoplasts are now avoided because they are labour intensive and the recovery of plants often requires long culture periods with the risk of generating somaclonal variants.

Protoplast Fusion

In plant breeding, the technique of somatic hybridization by protoplast fusion is a particularly useful way of introducing genetic variation from distant relatives into crop plants in circumstances where sexual hybridization is prevented by reproductive barriers. When two or more protoplasts fuse to form al new cell, the parent nuclei may remain separate or fuse to form a somatic hybrid. If one of the nuclei is lost after fusion, the cytoplasms of the two parent protoplasts will still coalesce and form what is known as a cytoplasmic hybrid or cybrid. Cytoplasmic hybrids contain the nucleus of one parent protoplast but a mixture of the cytoplasmic organelles of both parents. The production of cytoplasmic hybrids is an important technique in plant

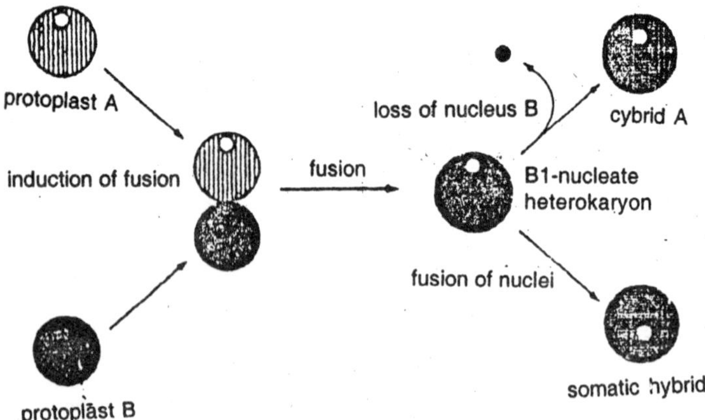

Fig. 7.4. Scheme showing the formation of somatic hybrids by protoplast fusion.

breeding programmes where CMS is desirable, because the CMS trait is inherited through mitochondria.

Isolated protoplasts can fuse spontaneously, although this occurs infrequently because of the mutual charge repulsion that results from the net negative surface charges on the protoplasts' plasma membranes. Nevertheless, fusion can be induced either by the application of compounds known as fusiogenic agents or by placing the protoplasts in an electric, field.

Fusiogenic agents tend to either neutralize or shield the surface charge so that adjacent protoplasts come in to close proximity, adhere and coalesce. Fusion is induced by high concentrations of calcium ions at high pH (pH 10) and high temperature (37°C). However, the most widely used fusiogenic agent is PEG.

Electrofusion takes place in two stages. First, the protoplast suspension is placed in a fusion cell between two metal electrodes and subjected to a non-uniform alternating electric field. The protoplast

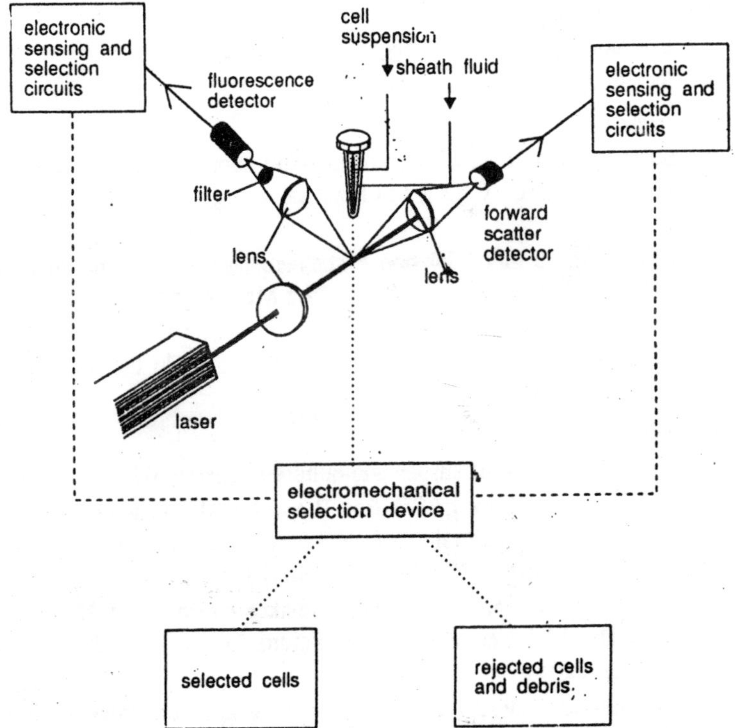

Fig. 7.5. Cell sorting by FACS (fluorescence activated cell sorting).

membranes become differentially charged and positive regions are attracted to negative regions on adjacent protoplasts resulting in the protoplasts aligning in chains. In the second step, a high-voltage DC pulse is applied to the aligned protoplasts that cause breakdown of adjacent membranes and fusion.

Flow cytometry can be used to identify the fusion products by cell sorting and by estimation of nuclear DNA.

PROTOCOL

Preparation of Protoplasts from Suspension Cultures (e.g. Maize) using a Long Incubation

Equipment

Sterile 250 ml conical flask

Bench centrifuge and centrifuge tubes

Pasteur pipette or automatic pipettor Light microscope

Orbital shaker

Aluminium foil

Epifluorescence microscope with UV filter set

Sterile 50 ml Falcon tubes.

Materials and reagents

Solution A: 100 ml of culture medium supplemented with 0.6 M (16.4 g) sorbitol, 0.43 g MS basal salts, 3 g sucrose, 2 mg l^{-1} 2,4-D, pH 5.2

Solution B: 25 ml of Solution A supplemented with enzyme mixture: 0.25 g Cellulase Onozuka RS, 0.25 g Macerozyme R10 (Onozuka) and 0.05 g pectolyase

Flask of suspension culture cells judged to be in the logarithmic phase of growth.

Procedure

1. Spin down 50 ml of suspension-cultured cells at 1000 g for 5 min.
2. Resuspend in 25 ml of Solution B and transfer to a sterile vessel that will allow maximum surface area (e.g. 30 ml of cell suspension in the bottom of a 250-ml conical flask).
3. Incubate at 50 rpm on an orbital shaker overnight at 25°C. The flask should be covered in aluminium foil to keep the enzyme solution dark.
4. Transfer to a sterile 50 ml Falcon tube and centrifuge at 600 g for 5 min.

5. Remove the supernatant with a pipette or a fine needle and discard.
6. Resuspend the protoplast pellet in 25 ml of Solution A using a wide-mouthed instrument such as a cut automatic pipette tip to reduce shear. Alternatively, add wash solution and partially invert tube very slowly.
7. Repeat centrifugation step once more. This should ensure that the enzymes are removed from the protoplasts. Finally, resuspend pellet in 5-10 ml of Solution A.
8. Make an initial check by light microscope for physical appearance (intact protoplasts are spherical) followed by FDA uptake using a fluorescence microscope. Cells that appear fluorescent green are viable.

Isolation of Protoplasts from Carrot Suspension Cultures Using a Rapid Incubation

Equipment

As above with the addition of sterile glass wool.

Materials and reagents

Flask of cells judged to be in logarithmic phase of growth

Solution A: 100 ml: 0.43 g MS medium, 2.5 g sucrose, 5 ml of coconut water, 100 $\mu g\ l^{-1}$ of 2.4-D, 100 $\mu g\ l^{-1}$ zeatin, 0.4 M (10.93 g) sorbitol, pH 5.0

Solution B, 25 ml: As solution A, but supplemented with 1 g Onozuka R10 cellulase, 0.5 g pectinase (Sigma).

Procedure

1. Take 50 ml of cells and centrifuge at 750 g for 10 min at 25°C.
2. Wash twice in 25 ml of Solution A and transfer to sterile 250 ml conical flask. Cover flask in aluminium foil to keep contents dark.
3. Shake on an orbital shaker at 50 rpm for 10 min, 25°C to ensure plasmolysis.
4. Repeat centrifuge step.
5. Resuspend pellet in 15 ml of Solution B and incubate on orbital shaker at 50 rpm in the dark at 25°C for 1-2 h.
6. Filter through sterile glass wool and rinse twice with 2-3 ml of Solution A.
7. To increase concentration of protoplasts, centrifuge at 600 g for 5 min and resuspend pellet in 5 ml of Solution A.
8. Use light microscope to examine preparation for intact protoplasts.
9. Assess small sample for viability using FDA.

Rapid Preparation of Mesophyll Protoplasts from Maize Leaves

Equipment

Razor blades
Forceps
Conical flask with a sidearm
Filter pump
Large Petri dish
Nylon sheet, 60 μm mesh
135 μm mesh spatula
Light microscope
Bench centrifuge.

Materials and reagents

Zea mays plants with well-developed leaves

Cellulase buffer (CB), 50 ml: 20 mM MES buffer pH 5.5 (0.43 g); 1 mM $MgCl_2$ (10.16 mg); 0.6 M sorbitol (8.2 g); 2% cellulase (1 g); 0.1% pectinase (0.05 g)

Wash buffer (WB), 100 ml: 50 mM Tris buffer (0.606 g) pH 8.0; 0.6 M sorbitol (16.4 g); 1 mM $MgCl_2$ (20.32 mg); 100 mM β-mercaptoethanol (0.701 ml).

Procedure

1. Take well-developed maize leaves and using the razor blade, slice into 0.5 1 mm strips.
2. Place strips in a vacuum flask containing 50 ml of cellulase buffer.
3. Apply vacuum using a filter pump. Maintain the negative pressure until all sections have become infiltrated with buffer.
4. Gently transfer to a large Petri dish and allow digestion to continue at room temperature for 3-5 h. Take a small volume (100 μl) to view under a light microscope to assess the level of digestion.
5. Filter the suspension through the 135-μm mesh to remove the broken cells and debris. Resuspend the partially digested fragments of leaf in 50 ml of WB in the original Petri dish.
6. Using the forceps, shake the leaf fragments in the WB to release the protoplasts. Agitating the Petri dish may also help.
7. Filter through the 60-μm mesh. Pellet the protoplasts by centrifugation at 300 g for 5 min. Gently resuspend in 25 ml of WB.
8. Repeat wash step and check viability of washed protoplasts with FDA.

Preparation of Mesophyll Protoplasts from Tobacco Leaves and Purification of Protoplasts on a Density Gradient

Equipment

Razor blades

50 ml of digestion medium for 10-15 g of plant tissue

Plastic Petri dish, 9 cm

Glass wool

Nylon mesh 100-200 μm pore size

Bench centrifuge, capable of operating at very low g-forces and centrifuge tubes

Glass Pasteur pipette

Light microscope

Tea strainer

Fine forceps.

Materials and reagents

Solution A, 50 ml: Digestion medium 500 mM D-sorbitol 4.56 g; 1 mM $CaCl_2$ 7.35 mg 5 mM MES-KOH, pH 5.5, 49 mg; Cellulase Onozuka R10, 1 g and Macerozyme R 10, 0.15 g

Solution B, 100 ml: Wash medium 500 mM D-sorbitol 9.11 g; 1 mM $CaCl_2$, 14.7 mg and 5 mM MES-KOH pH 6.0,98 mg

Solution C, 100 ml: density gradient solution 500 mM sucrose 8.56 g; 1 mM $CaCl_2$ 7.4 mg and 5 mM MES-KOH pH 6.0, 98 mg

Solution D, 100 ml: density gradient solution 400 mM sucrose 6.8 g; 100 mM D-sorbitol 0.90 g, 1 mM $CaCl_2$ 7.4 mg and 5 mM MES-KOH pH 6.0, 98 mg

Tobacco plant with well-developed leaves, 3-4 weeks old.

Procedure

1. Take a leaf and gently score underside with a sharp razor blade until the entire area is criss-crossed with fine, parallel cuts. Avoid cutting completely through the upper epidermis. Lay the cut area down on the surface of digestion medium in a Petri dish. Repeat this procedure until the surface of the Petri dish is covered.
2. Replace Petri dish lid and incubate in the dark for 16 h at room temperature. Repeat this, using fresh Petri dishes until there is sufficient material for your purposes.
3. Using fine forceps to hold the stem, shake the leaf on the surface of the digestion medium. The medium should change colour to green as protoplasts leave leaf structure.

4. Collect the protoplast suspensions from each Petri dish and pool.
5. Pour through a tea strainer and then the nylon mesh to remove cell debris. Wash each membrane gently after use to remove protoplasts.
6. Centrifuge combined washes at 50-100 g for 3 min. The protoplasts should collect at the bottom of the centrifuge tube. Discard the supernatant.
7. Gently resuspend protoplast pellet in 10 ml of Solution C and aliquot 5 ml to each of two centrifuge tubes.
8. Slowly add 5 ml of Solution D to each centrifuge tube, then add 5 ml of Solution B to make a three step gradient These solutions should be added gently while the tube is slanted so that they run slowly down the side of the tube and generates discrete steps.
9. Centrifuge at 300 g for 5 min.
10. The protoplasts will appear at the interface between the top two layers. Carefully remove the top layer with a pipette and then collect the protoplasts.
11. Check this preparation under the light microscope. Avoid putting a coverslip on the slide as it can damage the fragile protoplasts. It should be possible to see that the protoplasts are intact and not osmotically stressed. Stressed protoplasts can be identified by the cytoplasm appearing to be dense and compacted. Healthy protoplasts should also be perfectly spherical.

Rapid Preparation of Protoplasts from Maize Roots

Equipment

Scalpel

Shaking water bath

Glass wool

Bench centrifuge capable of very low speeds and tubes

Parafilm

Petri dish

100-ml Erlenmeyer flask

Materials and reagents

Protoplasting medium (PM), 50 ml: 5 mM MES-KOH (49 mg), pH 6.0; 0.5 M sorbitol (4.56 g), 1 mM $CaCl_2$ (7.35 mg); 0.5% BSA (0.25 g), 0.8% cellulase (Onozuka R10) (0.4 g), and 0.08% pectolyase (0.04 ml)

Wash medium (WM), 200 ml: 5 mM MES-KOH pH 6.0 (196 mg); 0.5 M sucrose (8.5 g) and 1 mM $CaCl_2$ (29.4 mg)

Density gradient step 1 (DG1), 50 ml: 5 mM MES-KOH (49 mg), pH 6.0; 0, 1 M sorbitol (0.91 g); 0.4 M sucrose (3.4 g) and 1 mM $CaC1_2$ (7.35 mg)

Density gradient step 2 (DG2), 50 ml: 5 mM MES-KOH (49 mg), pH 6.0; 0.5 M sorbiool (4.56 g), 1 mM $CaC1_2$ (7.35 mg)

Maize seedlings grown hydroponically with 8-10 cm roots.

Procedure

1. Take roots and excise 8 cm from root tip, excluding any shoot material.
2. Using a sharp razor blade finely chop 20 g of root material in 20 ml of protoplasting medium in a Petri dish. The tissue should be sliced to a fine pulp.
3. Transfer the homogenate to a 100 ml Erlenmeyer flask and incubate at 28°C in the dark for 3 h in a shaking water bath.
4. Filter the suspension through glass wool.
5. Centrifuge the filtrate at 60 g for 5 min.
6. Resuspend the pellet in 5 ml of ice-cold wash medium.
7. Layer on to this first 2 ml of DG1, then 1 ml of DG2 to form a step gradient.
8. Centrifuge this at 200 g for 5 min. The protoplasts will form a layer on the interface between the top two layers.
9. Wash protoplasts in 5 ml of DG2 and centrifuge at 100 g for 10 min. Resuspend pellet in 1 ml of DG2.
10. Assess for viability using FDA and keep on ice until ready to use.

Preparation of Protoplasts from Roots (Arabidopsis) with Partial Purification of Protoplasts by Flotation

Equipment

Sterile 250 ml Erlenmeyer flask

0.22 μm filters to sterilize heat-labile solutions

Autoclave

pH meter

Sterile Petri dishes

Growth chamber with orbital shaker

Scalpel, scissors and forceps

Broad-mouthed pipette

100-, 50- and 25-μm sieves or meshes

Bench centrifuge capable of very low speeds and centrifuge tubes

Haemocytometer.

Materials and reagents

Sterile distilled water

10% v/v sodium hypochlorite containing 0.05 ml l^{-1} Tween 20

Basal medium (BM): 4.3 g l^{-1} of MS medium, 3% sucrose, B5 vitamins (Sigma), pH 5.8

0.5 BM agar plates (2.2 g l^{-1} MS, 3% sucrose, B5 vitamins, pH 5-8) 0.8% agar

50 ml aliquots of 0.5 BM in 250 ml Erlenmeyer flasks, sterile

MSAR 1 medium: BM plus 2 mg l^{-1} 1AA, 0.5 mg l^{-1} 2,4-D, 0.5 mg l^{-1} 6-(γ,γ-dimethylallylamino) purine riboside (IPAR)

0.45 M sucrose solution

Protoplasting medium (PM): 0.5 BM containing 0.45 M sucrose or 0.45 M mannitol

Enzyme solution: 1% cellulase (Onozuka R10), 0.25% macerozyme (R10 Serva) dissolved in PM

0.45 M mannicol

Arabidopsis roots grown aseptically on agar.

Procedure

1. Grow arabidopsis seeds (var. Columbia or C 24) on 0.5 BM plates for 10-14 days in a light regime of 16 h light and 8 h dark at 22°C.
2. Aseptically remove plantlets from plate and separate the roots from the green tissue. Chop the roots into 2-4-mm pieces, using the sterile scalpel and transfer into Petri dishes with 10 ml of MSAR 1 medium.
3. Incubate in the dark or under low light conditions, shaking at 100 rpm for 7-12 days.
4. Remove the medium and wash the root explants once with 0.45 M sucrose solution. Replace with 20 ml of enzyme solution
5. Incubate for 12-16 h with occasional shaking.
6. Collect the protoplasts with a broad-mouthed pipette and pass through the 100-, 50- and 25-μm sieves.
7. Centrifuge the suspension for 5 min at 50 rpm.

8. Collect the band of floating protoplasts concentrated at the top of the solution and transfer to a new tube. At this stage, it may be necessary to pool the contents of two or more tubes to obtain a working number of protoplasts for the next steps.
9. Resuspend in 0.45 M mannitol and pellet at 60 g for 5 min. Repeat this step twice to completely remove the enzymes.
10. Check for viability using FDA.

Culture of Maize Protoplasts using Nurse Culture

Equipment

Autoclave

Tissue culture dishes (15 mm)

Balance

Sterile pipettes (5 ml) and pipetting device

Automatic pipette and sterile tips

Bench centrifuge capable of very low speeds and tubes

pH meter

Sterile 0.8 μm Millipore filters

Parafilm

Sterile forceps

Haemocytometer

Light microscope.

Materials and reagents

(Solutions should be sterilized)

Stock solution of 2,4-D at 1 mg/10 ml of ethanol : water 1:1

Stock solution of thiamine-HCl at 50 mg l^{-1} of H_2O (20 ml)

Modified White Vitamins: glycine 200 mg l^{-1}, nicotinic acid 50 mg l^{-1}, pyridoxine HCl 50 mg l^{-1}, thiamine HCl 50 mg l^{-1}

Protoplast culture medium (PCM) 100 ml (= 4 × 9 cm Petri dishes): 0.43 g of MS salts, 2 ml of 2,4-D stock solution, 0.025 g of glucose, 1 ml of stock thiamine, 2 g of sucrose, 2 ml of coconut water, 3.64 g of mannitol, pH to 5.9 with 0.1 M KOH PCMLMP: PCM medium with 0.8 g of low-melting-point (LMP) agarose

Postfeeder culture medium-MS 2D 100 ml: 0.43 g MS salts, 1 ml of Modified White Vitamins, 10 mg of myoinositol, 2 g of sucrose, 2 ml of stock 2,4-D, pH to 5.9 with 0.1 M KOH, 0.4 g of LMP agarose

Maize protoplasts

Procedure

1. Prepare nurse cultures. These can be prepared during the protoplast digestion. The media should be prepared and autoclaved for 15 min at 121°C.
2. Aseptically pipette 3 ml of PCM-LMP into sterile, 60 × 15 mm culture dishes and allow to set.
3. Add an additional 3 ml of PCM-LMP containing 0.2 ml PCV of feeder cells from original cell suspension culture. PCM-LMP should be cooled to 40-45°C before mixing with the feeder cells.
4. Take 0.01 ml of the protoplast suspension and count. Adjust the concentration to 1 × 10^6 protoplasts ml^{-1} of PCM.
5. Plate 0.2 ml of the suspension on to the 0.8-μm filters resting on top of the feeder layer. Ensure that the plates are not too dry.
6. Leave the closed feeder plates in the laminar flow cabinet to let the protoplasts settle on to the filters. This may take from several hours to overnight.
7. Seal plates with Parafilm and incubate at 25°C in the dark.
8. Transfer filters carrying protoplasts to fresh feeder plates at weekly intervals using sterile forceps.
9. After 2-3 weeks„ callus should be visible to the naked eye and can be moved using sterile tools to agar plates.

Culture of Arabidopsis Protoplasts using the Hanging Drop Method

Equipment

55-mm Petri dishes
Sterile wide-bore pipette
Sterile spatula
Growth chamber
Light microscope
Haemocytometer.

Materials and reagents

(Solutions should be sterilized)
Flask of freshly prepared protoplasts
Basal medium (BM): 4.3 g l^{-1} of MS basal salts, B5 vitamins
Sodium alginate solution: 1% w/v solution of sodium alginate in BM medium containing 0.45 M sucrose
Calcium-agar plates: 20 mM $CaCl_2$, 0.45 M sucrose and 1% agar
PM, protoplast medium: 2.21 g l^{-1} MS basal salts containing 0.45

M sucrose, MSAR 1 hormones (2 mg l^{-1} IAA, 0.5 mg l^{-1} 2,4-D, 0.5 mg l^{-1} IPAR

MSAR 1 medium: 4.3 g l^{-1} MS plus 2 mg l^{-1} IAA, 0.5 mg l^{-1} 2,4-D, 0.5 mg l^{-1} IPAR.

Procedure

1. Suspend protoplast pellet from final wash in 1 ml of sodium alginate solution and count protoplast yield using haemocytometer. Adjust density to 5-7 $\times$ 10^5 cells ml^{-1} using more sodium alginate if necessary.
2. Using a wide-bore pipette, transfer the protoplast-sodium alginate suspension as 250-500-μl drops on the calcium agar plates.
3. Allow the alginate to form a gel for 45 min.
4. Transfer the individual droplets with a spatula into 55-mm Petri dishes containing 5 ml of PM and culture in a growth chamber at 22°C under dim light.
5. Remove 2.5 ml of PM and replace with 2.5 ml of fresh PM on days 7 and 14.
6. Remove 1 ml of PM and replace with 1 ml of MSAR1 medium on days 21, 28 and 35.
7. By day 35, microcalli should be visible.

Note

This protocol can also be applied to protoplasts derived from roots and cell suspension cultures. However, for initiating liquid cultures of protoplasts, the density needs to be adjusted to 1 $\times$ 10^6 cells ml^{-1} until cell divisions start when it should be diluted with PM to 5 $\times$ 10^5 cells ml^{-1}.

Detecting Regeneration of Cell Wall in Isolated Protoplasts

Equipment

Epifluorescence microscope with BG 12 filter set

Slides and coverslips

Automatic pipetting device and tips.

Materials and reagents

Calcofluor White medium (CWM) 0.05% Calcofluor White in culture medium supplemented with 0.4 M sorbitol

Protoplast suspension.

Procedure

1. Resuspend a small volume of protoplasts in CWM medium and observe under an epifluorescence microscope. Calcofluor White

stains only cell wall material and exhibits fluorescence when irradiated with blue light. As the cell wall is unlikely to reappear until the protoplasts have been in the appropriate regenerating medium for a minimum of 2 h, the Calcofluor White staining will not appear until this time has elapsed.

2. Initially, the staining will manifest itself as small, fluorescent dots on the surface of a small percentage of the population. After 5-6 h has elapsed, more protoplasts will show this pattern of staining.
3. If the treatment is continued for 18-20 h, large areas of dense staining will appear on viable, healthy protoplasts. Eventually, the entire cell surface will be covered with fluorescent staining as the cell wall is regenerated completely.

Protoplast Fusion Induced by Polyethylene Glycol (PEG)

Equipment

Bench centrifuge capable of very low speed (50 g) and tubes

Pipette

Timer

Parafilm

Petri dishes (9 cm).

Materials and reagents

(Solutions should be sterilized)

PEG fusion solution: 30% w/v PEG 6000, 4% w/v sucrose, 10 mM $CaCl_2$. This should be autoclaved and stored in dark until used.

Protoplast maintenance medium appropriate to the types of protoplasts to be used. Also to be used as a washing medium. These should be sterile.

9-cm Petri dishes containing 8 ml of sterile protoplast medium plus 0.8% agarose.

Procedure

1. Ensure that the densities of the two protoplast populations are both 2×10^5 cells ml^{-1} .
2. Spin 4 ml of each population together in one tube, at 100 g for 10 min to pellet the protoplasts.
3. Remove the surplus medium so as to leave the pellet in 0.5 ml of medium.
4. Add 2 ml of PEG fusion solution to each tube and leave for 10 min.

5. Sequentially dilute the fusion mixture at 5 min intervals, by adding 0.5, 1, 2, 2, 3 and finally 4 ml of protoplast maintenance medium per tube.
6. Gently mix tube after each addition.
7. Pellet protoplasts at 100 g for 10 min and remove supernatant.
8. Wash the protoplasts once in protoplast medium.
9. Resuspend finally in 16 ml of protoplast medium.
10. Aliquot 8 ml, using a wide-mouthed pipette, onto the surface of the agarose plates.
11. Seal the plates with Parafilm and culture at 25°C under a low light regime.
12. Microcalli should be visible after several weeks. Blocks of agarose may be aseptically cut out and transferred to liquid culture if required.

Electrical Fusion of Protoplasts

Equipment

Electrofusion apparatus including sterile fusion chambers and electrodes.

Materials and reagents

(These media should be sterile)

Protoplast maintenance solutions suitable for both populations

Solid media suitable for post-electrofusion protoplasts, details of precise media for the species chosen should be collected from the literature.

Procedure Part A—Electrofusion

1. The suspension of mixed protoplast populations should be placed in the electrofusion chamber. A weak AC current (400,000 Hz, 1.5 V) is passed between the electrodes, through the protoplasts, causing them to become positively charged on one side and negatively charged on the other. The protoplasts align themselves in chains. Varying protoplast density, frequency of AC field and peak-to-peak voltage can alter the numbers in each chain.
2. After a period of approximately 90 s, the AC field is replaced by a single, high. voltage pulse discharge of, for example, 1000 V cm^{-1}. Where the protoplasts are in contact with each other the plasma membranes break and fuse forming a continuous membrane around the pearl chains. This is followed by cytoplasmic fusion.

Multiple fusions will occur at high concentrations of protoplasts but this can be adjusted with experience.

3. The protoplast suspension is then plated out and cultured or as in the literature for the species used. Electrofusion can be less productive than PEG fusion.

Procedure Part B—Selection of heterokaryons

After chemical or electrical fusion of protoplasts, there will be a mixture of unfused protoplasts, fused parents of the same origin, multiple fusions and the desired heterokaryons. Isolation of the correct product can be achieved by a choice of the methods below.

1. The simplest technique is to allow all products of fusion to mature and regenerate and then to identify the heterokaryons by morphological differences at the seedling stage.
2. If the two parent populations have obviously different visual properties, it may be possible to separate fused heterokaryons using a microscope and a micromanipulator. This procedure may be laborious in practice.
3. If the parents are not easily distinguishable by sight, the parent populations may be stained with two different colour dyes. For example, FDA fluoresces yellow-green while rhodamine isothiocyanate is red. Selection of fused protoplasts showing dual fluorescence allows identification of heterokaryons.
4. A flow cytometer equipped with a UV source can be calibrated to separate dual-labelled protoplasts from single-labelled. Flow cytometry can process a large number of protoplasts accurately.

Protoplast Transformation by Electroporation

Equipment

All equipment required to prepare protoplasts

Small Petri dishes

Ice and Ice bucket

Electroporation apparatus with 1.0 ml cuvettes containing built-in electrodes

Timer.

Materials and reagents

(Solutions should be sterilized)

Electroporation buffer (EB): 1 mM MES, pH 5.8, 2.5 mM $CaCl_2$, 0.225 M sorbitol, 0.225 M mannitol

MS medium: appropriate culture medium supplemented with 0.6 M sorbitol.

Procedure

1. Resuspend protoplast pellet gently in 25 ml of EB. Spin at 600 g for 10 min.
2. Repeat protoplast wash 4-5 times with 25-50 ml of EB.
3. Resuspend final pellet in 5 ml of EB.
4. Add DNA in a ratio of 100 μl of DNA : 500 μl cells.
5. Aliquot 600 μl volumes into Petri dishes for electroporation and leave on ice for 5 min.
6. Electroporate, following manufacturer's instructions and recommendations. Typical conditions recommended are: capacitance 910 μF, voltage 120-140 V, time of discharge 1000 ms, volume 600 μl and electrode distance is 0.2 mm.
7. Rest the electroporated protoplasts on ice for 15 min.
8. Collect the electroporated protoplasts in a Petri dish. Add 0.9 ml of EB and 1.6 ml of MS medium.
9. Incubate in dark at 22°C. Expression should be apparent after 12 h.

Transformation of Protoplasts Mediated by Polyethylene Glycol (PEG)

Equipment

Ice and ice bucket

Bench centrifuge capable of low speed and tubes

Haemocytometer and microscope

Glass Petri dish (9 cm)

Automatic pipetting device and tips.

Materials and reagents

All solutions needed to prepare protoplasts. These need to be sterile

0.5 M MaMg solution; 0.5 M mannitol, 15 mM $MgCl_2.6H_2O$, 0.2% MES, pH 5.8 (sterile)

PEG solution: 40% w/v PEG 1450 in MaMg solution. This should be filter-sterilized

Plasmid DNA solution in water.

Procedure

1. Resuspend protoplast pellet in 1 ml of MaMg solution.
2. Count cells and adjust density to approximately 1×10^6 cells ml^{-1}.

3. Place protoplast suspension on ice for 35 min.
4. Centrifuge protoplasts at 60 g for 5 min.
5. Resuspend pellet in 0.3 ml of MaMg solution and carefully transfer them as single droplets to the middle of a 9 cm glass Petri dish.
6. Slowly add 25-35 μg of plasmid DNA dissolved in water into the drop of protoplast suspension. Rock the Petri dish very gently to mix the protoplasts and the DNA.
7. After 5 min, add 0.3 ml of PEG solution at the circumference of the drop of protoplast suspension.
8. Again tilt the Petri dish very gently to mix the PEG and the protoplasts. Alternatively, mix with a sterile micropipette tip.
9. After 10 min, add 1 ml of 0.45 M mannitol solution to the sides of the droplet.
10. At 2-min intervals, add 2 ml of 0.45 M mannitol solution with gentle shaking until about 12 ml are present in the Petri dish.
11. Collect the suspension by wide-mouthed pipette and transfer to a centrifuge tube. Centrifuge at 60 g for 5 min.
12. At this stage, the protoplasts can be studied as a transient expression system or transferred to selective culture medium to regenerate stably transformed calli or plantlets. If the protoplasts are to be subcultured, the entire procedure should be carried out under aseptic conditions.

8

HAPLOID TISSUE CULTURE

One of the most significant biotechnological advances in commercial plant breeding in recent years has been the development of *in vitro* techniques to produce haploid plants that can be used to generate homozygous lines.

In traditional plant breeding programmes, crosses between distantly related species can bring about novel gene combinations. However, the hybrid offspring can be few in number and may be genetically unstable. More importantly, these hybrids require several generations (6-10) of further crossing or selfing to attain the degree of homozygosity that is necessary to ensure that advantageous characteristics are fixed and to eliminate any undesirable recessive traits. Alternative procedures for the production of homozygous lines in one or two generations can therefore save several years of conventional plant breeding. The formation of dihaploids from haploid plants is a quick and valuable way of producing homozygous lines from heterozygous parents.

Haploid plants have a simple or gametophytic set (1*n*) of chromosomes and may be sterile. However haploid plants can double their chromosome set spontaneously or this can be induced by chemical agents (e.g. colchicine, the mitotic spindle disrupter, results in chromosome duplication without cell division) to produce fertile doubled haploids (dihaploids). An advantage of this procedure is that the resulting dihaploids are homozygous for all genes and therefore true-breeding, making the selection process for desirable traits mote efficient.

In theory, haploid plants could be produced by the *in vitro* culture of haploid gametophytic tissues, that is, unpollinated ovaries or ovules and anthers or pollen. Viable haploid plants have been produced by ovary and ovule culture for some crops; the procedures are however technically laborious and the yields low. Where are two *in vitro* procedures that are now routinely used in the production of haploids: (i) anther culture and (ii) the culture of isolated microspores. The development of haploid plants from male gametophytic tissue is known as androgenesis.

Anther Culture

Anther culture is a rapid and relatively simple procedure by which pollen-producing anthers are collected from flowers at a certain stage of development; sometimes they are given a low- or high-temperature pretreatment, before sterilization and culturing on a specific nutrient medium. Mature pollen (the male gametophyte) develops from haploid microspores that are in turn formed by meiosis in pollen mother cells. Microspores have a single nucleus and mature pollen three nuclei. When anthers containing microspores at the middle to late uninucleate stage are cultured on solid media, individual microspores form callus. Plants with half the chromosome number can be regenerated from the callus either directly by embryogenesis or indirectly by organogenesis. Alternatively, microspores may give rise to embryoids (aggregates of cells with recognizable embryonic structures) that are capable of differentiation into embryos.

Microspore Culture

Microspore culture involves the physical isolation of microspores from anthers at a specific stage of development. After a suitable pre-treatment, the anthers are homogenized and microspores are isolated and purified from the homogenate by filtration and centrifugation steps. The purified microspores are then cultured in a liquid medium. Small embryos develop directly from the microspores (microspore embryogenesis) and are transferred to a regeneration medium for plantlet development.

Anther Culture versus Microspore Culture

Androgenesis has been used to produce haploids from several hundred different plant species including commercially important cultivars. The choice of which procedure to adopt for haploid production is determined by a number of considerations. Anther culture is a relatively simple and rapid technique that can produce large numbers of haploids with minimal equipment. The major problems generally encountered in the use of anther culture are genotype dependency and the high frequency of albino plants and mixed ploidy types (mixoploids)

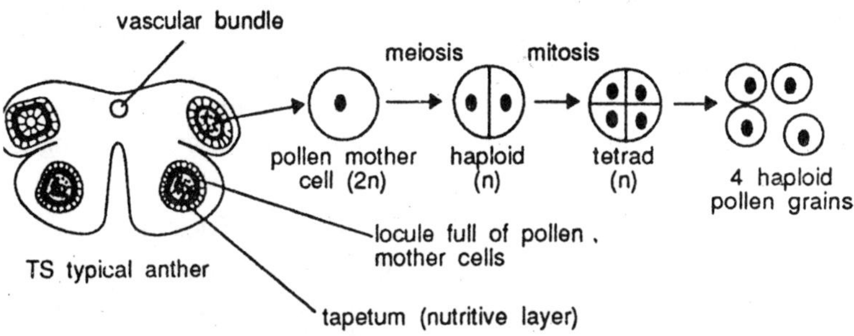

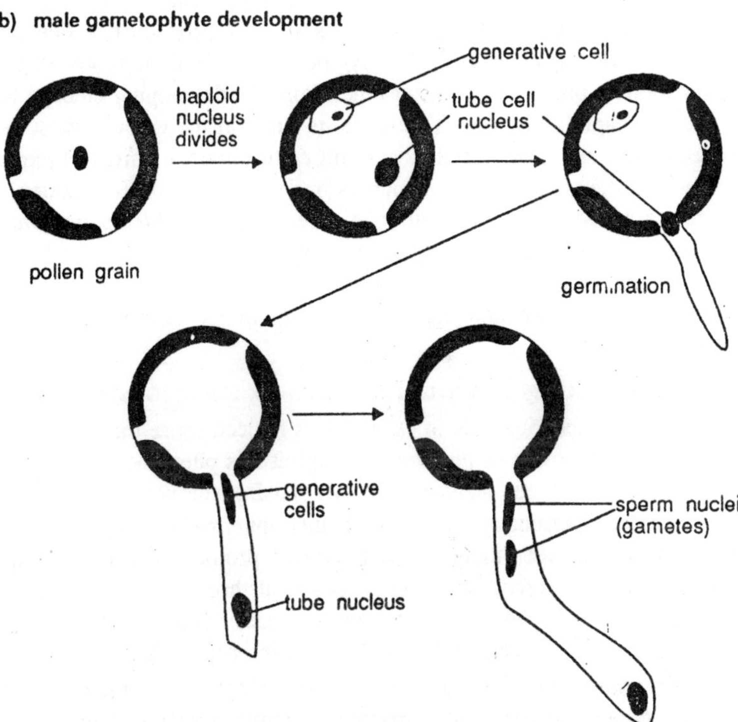

Fig. 8.1. Pollen development.

in the regenerated populations. In comparison to anther culture, the microspore technique is technically more complicated and requires more extensive laboratory facilities. However, in species where experimental comparisons have been made, problems such as mixoploidy are less frequent in populations derived from microspore cultures than anther cultures. The use of microspore cultures also has other advantages: (i) development of embryos directly from microspores without an intermediate callus stage reduces the risk of genetic abnormality; (ii) microspore cultures are ideal material for genetic transformation, as insertion of foreign DNA into the haploid genome will give rise to homozygous expression after chromosome doubling to form the dihaploid.

Intergeneric Crosses and Embryo Rescue

An alternative method for rapid haploid production involves intergeneric crossing followed by embryo rescue. Such wide crosses may be incompatible and viable seed development may not occur. However, a haploid embryo may form and this can be rescued by excision and tissue culture. The basis of this procedure relies on fertilization to produce a diploid zygote, which is then followed by elimination of one set of chromosomes from the developing embryo to produce a haploid embryo. To ensure its survival the embryo must be dissected out and transferred to nutrient culture media. This technique has been successful in cereals such as barley and wheat. For example in the so-called *bulbosum* technique, bread wheat *Triticum aestivum* (female parent) is crossed with the barley *Hordeum bulbosum* to yield haploid plants. Other wide crosses include bread wheat and durum wheat *Triticum turgidum* spp. *durum* with maize *Zea mays* as the paternal parent.

Procedures for the Induction of Androgenesis

Although androgenesis in cereals was noticed more than 70 years ago, the development of protocols to exploit this phenomenon in plant breeding has only been achieved recently. Androgenesis has been tested in several economically important plants, but many crops are still recalcitrant. As yet, there is no universal protocol common to all plant species. In general, protocols consist of three sequential phases. The protocol for barley microspore embryogenesis is a typical example. In this case, the procedure is divided into three principal steps:

1. *Pre-treatment of anthers*—anthers at the right developmental stage are given a stress pre-treatment for approximately 4 days.
2. *Microspore culture*—microspores are isolated from the treated anthers and put into culture. Cell division can be observed after 4 days and multicellular structures are formed after 14 days.
3. Development, of embryos—multicellular structures develop into embryos and further growth into plantlets requires an additional 21 days.

There are many variations of this basic protocol. Successful microspore regeneration depends on the plant material. The variety, age and the growth status of the plant are all important factors. The role of the pretreatment stress is to induce the microspores to switch from the gametophytic developmental pathway to sporophytic development. The pretreatment may be temperature (cold or heat shock), starvation or osmotic stress, and can be applied to intact flowers (e.g. barley spikes), isolated anthers or isolated microspores. The stress hormone abscisic acid is implicated in the reprogramming of microspore development.

During the pre-treatment, some of the microspores develop into embryogenic microspores that can develop into embryos. Embryogenic microspores are characteristically twice the size of non-embrogenic microsporess. After pre-treatment, the microspores are cultured in a specific medium where cell division and differentiation occur. At this stage, chromosome doubling to form dihaploids may occur spontaneously or can be induced by agents such as colchicine

Protocol

Anther Culture of Wheat (*Triticum aestivum* L.)

Equipment

Sterile dissecting instruments

Dissecting microscope

Incubator, 28°C.

Materials and reagents

Greenhquse grown wheat plants

Petri dishes (600 × 14 mm) containing 4.5 ml of CHB-2 medium [KNO_3, 1415 mg l^{-1} ; $(NH_4)_2SO4$, 232 mg l^{-1} , KH_2PO_4, 200 mg l^{-1} ; $CaC1_2.2H_2O$, 83 mg l^{-1} ; $MgSO_4.7H_2O$, 93 mg l^{-1}, $FeNa_2$ EDTA, 32 mg l^{-1}; $ZnSO_4.7H_2O$, 5 mg l^{-1}; $MnSO_4.4H_2O$, 5 mg l^{-1}; H_3BO_3, 5 mg l^{-1}; Kl, 0.4 mg l^{-1}; $Na_2MoO_4.2H_2O$, 0.0125 mg l^{-1}; $CuSO_4.H_2O$, 0.0125 mg l^{-1}; $CoC1_26H_2O$, 0.0125 mg l^{-1}; glycine 1.0 mg l^{-1}; thiamine-HCl, 2.5 mg l^{-1}; Pyridoxine-HCl, 0.5 mg l^{-1}, nicotinic acid, 0.5 mg l^{-1}; biotin, 0.25 mg l^{-1}; calcium pantothenate, 0.25 mg l^{-1}; ascorbic acid, 0.5 mg l^{-1}; myo-inositol, 300 mg l^{-1}; glutamine, 1000 mg l^{-1}; glucose, 0.21 mg l^{-1}; 2,4-D, 0.5 mg 1^{-1}; kinetin. 0.5 mg l^{-1}; pH 4.5, supplemented with 9% (w/v) maltose (sterile).

70% Ethanol spray

Sodium hypochlorite (20%)

Acetocarmine solution (commercial preparation)

Sterile distilled water

Sterile filter paper.

Procedure

1. Check microspore development in anthers from spikes just prior to emergence of first leaf blade. Remove anthers, stain with acetocarmine and examine with a microscope.
2. Collect spikes containing anthers with microspores in the mid- to late-uninucleate stage.
3. Store spikes with their cut ends in tap water at 5°C for 5 days.
4. Spray spikes with 70% alcohol, and surface sterilize with sodium hypochlorite for 20 min.
5. Wash spikes three times with sterile water and blot dry with sterile paper tissue.
6. Dissect out the anthers under sterile conditions and add anthers from one spike to each Petri dish containing CHB-2 medium.
7. Incubate cultures in the dark at 28°C for 35 days.
8. For plant regeneration—at the end of the incubation period, remove embryos and transfer to 190.2 medium solidified with 3.5 g l^{-1} Gelrite. Incubate at 25°C and with a light intensity of 60-80 μmol m^{-2} s^{-1}.

Wheat Microspore Culture

Equipment

Blender (Waring or similar, container sterilized)

Stainless steel mesh 100-μm (sterilized)

Bench centrifuge with swing-out rotor and sterile centrifuge tubes

Haemocytometer

Incubator 28°C

Microscope (ultraviolet).

Materials and reagents

Materials and reagents as for previous Protocol

Mannitol 0.3 M (sterile)

Maltose (21% w/v) (sterile)

Petri dishes (35 × 10 mm)

Fluorescein diacetate.

Procedure

1. Grow plants and collect and treat spikes.
2. Remove spikelets from five spikes and bend with a Waring blender in 0.3 M mannitol.
3. Filter slurry through the stainless steel mesh.
4. Centrifuge filtrate at 55 g for 5 min to pellet microspores.
5. Resuspend microspore pellet in 0.3 M mannitol.
6. Layer suspension on top of a 21% (w/v) maltose solution in a centrifuge tube.
7. Centrifuge at 55 g for 5 min. Collect microspores accumulated at gradient-interface.
8. Wash microspore fraction once by resuspension in mannitol and pelleting at 55 g for 5 min.
9. Check viability of microspores with FDA and determine density with a haemocytometer.
10. Resuspend in CHB-2 medium with 9% (w/v) maltose at a concentration of 50000 microspores ml^{-1}.
11. Transfer 2.0 ml of microspore suspension to 35 ×10 mm Petri dishes, incubate at 28°C for 28 days.
12. For plant regeneration—at the end of the incubation period, remove embryos and transfer to 190-2 medium solidified with 3.5 g l^{-1} Gelrite. Incubate at 25°C and with a light intensity of 60-80 $\mu mol\ m^{-2}\ s^{-1}$.

9

ORGAN AND EMBRYO CULTURE

Maintaining and growing plant organs in culture has a variety of applications, from the experimental understanding of plant development to their use for the production of high-value products. An important feature of organ culture is that the organ grows and develops in a manner similar to that of the parent plant; this distinguishes them from callus or suspension-cultures (cell and tissue cultures).

A wide range of plant organs has been successfully cultured. The earliest experiments undertaken by White (1934) showed that root explants cultured in a liquid medium with inorganic nutrients, sucrose and yeast extract continued to grow at a rate of about 10 mm day^{-1}. Lateral roots developed after 4 days and could be used to initiate new root cultures. Protocols for root cultures will be described later in this chapter. Subsequently, leaves, shoot tips, complete flowers, anthers, pollen, ovules and intact fruit have all been successfully cultured. A protocol for maize ear culture is given as an example of the culture of an entire organ.

HAIRY ROOTS

Hairy root cultures are generated when a plant tissue is transformed with a culture of the bacterium *Agrobacterium rhizogenes*. The method of transformation is similar to that of *A. tumefaciens*, which transfers a segment of plasmid DNA to the cells of an infected plant. The transformation results in altered hormone levels, causing the induction, of roots and a root system that branches much more frequently than the usual root system of that plant, and is covered with a mass of tiny root hairs. This root system may be maintained indefinitely in liquid culture without the parent plant. Importantly; for cell and tissue culture, the transformed roots produce secondary metabolites at levels similar to those of the original plant. Hairy root cultures are a specialized form of organ culture and can be maintained continuously in bioreactors. It is based on the method of Jones and Sutton (1997), and could be adapted for hairy root cultures of other dicots.

EMBRYO CULTURE

Embryo culture, where the embryo is derived from the seed of a parent plant, is an important method both for the study of seed and embryo development, and as a means of obtaining viable seedlings when germination is a problem. The easiest method of culturing an embryo is to remove it from the seed when nearly mature and ready to germinate. However, in other instances (for instance to obtain sterile seedlings or to rescue mutant embryos), it may be necessary to dissect out an immature embryo and grow it directly on solid medium. The nutrient requirements of embryos that have been removed from the ovary is much more complex. Supplementing the culture medium, with coconut water, a plant liquid endosperm was shown by van Overbeek in 1941 to be successful in providing the nutrients and growth factors required. As embryos mature, they become increasingly autotrophic, and more complex media are not required. A variety of media have been used for embryo culture. In addition to a basal medium (e.g. Gamborg's salts), sucrose or glucose is the usual carbohydrate source; the amino acid glutamine is preffered as nitrogen source in comparison with asparagine. Other additions include hormones (particularly auxins and cytokinins) and undefined additions like coconut milk and casein hydrolysate. The concentration of agar has been found to be of particular importance as too high a concentration leads to dehydration of the embryo.

Embryo rescue techniques are important where a high rate of spontaneous abortion of embryos is likely to occur, for instance in the production of hybrids between closely related species, or in the production of transgenic plants. In the technique, embryos are isolated and grown on solid media as described above for embryo culture. The method described below for embryo rescue is for interspecific hybrids of *Phaseolus vulgaris* and *Phaseolus polyanthus*. It involves the removal of immature (early heart-shaped) embryos from seeds and their culture on defined media based on those of Gamborg or MS, supplemented with vitamins, a nitrogen source, the cytokinin BAP with sucrose as carbon source and 0.8% agar as gelling agent. After 20 days, the embryos are transferred to a rooting medium (with GA_3) before being established after a further 15 days in the glasshouse.

MINITUBERS AND MICROTUBERS

Organ culture techniques are important in commercial potato growing where disease-free tubers are required for propagation. Minitubers are produced *in vitro* by plantlets from meristem tip culture in sterile culture. Plantlets develop in 4-8 weeks from a meristem explant and are routinely subcultured by nodal cuttings. Rooted cuttings are then used to generate minitubers—small seed tubers, less than 3 cm in diameter in sterile media. The plants that produce the minitubers are tested extensively for major potato pathogens. Minitubers are stored in a controlled atmosphere until needed, when sprouting is induced. Alternatively, microtubers, which are up to 5 mm diameter, are produced in *in vitro* culture in the axils of the leaves. Microtubers can be either planted in the greenhouse to produce plantlets or minitubers, or directly planted into the field.

PROTOCOL

Isolation and Culture of the Primary Seedling Root of Dicots

Equipment

3 × 250 ml conical flasks containing 200 ml of sterile distilled water

24 × 90 mm Petri dishes

12 × glass culture tubes (150 × 25 mm) containing 20 ml of sterile distilled water

12 × glass culture tubes (150 × 25 mm) containing 20 ml of rooting medium (MR) culture medium

3 × scalpels

2 × pairs forceps.

Materials and reagents

MR: $Ca(NO_3)_2.4H_2O$, 236 mg l^{-1}; $FeSO_4.7H_2O$, 2 mg l^{-1}; KH_2PO_4. 12 mg l^{-1}; KNO_3, 81 mg l^{-1}; KCl, 65 mg l^{-1}; $MgSO_4.7H_2O$, 36 mg l^{-1}; sucrose, 4%; pH 5.8

24 pea seeds

250 ml conical flask containing 200 ml of 10% w/v calcium hypochlorite

1 × Pyrex conical flask containing 100 ml of 95% ethanol

2 × pieces of muslin (60 × 60-mm)

1 × 250 ml conical flask containing 200 ml of 1% aq. Tween 20.

Procedure

1. Wrap seeds in muslin and immerse in 95% ethanol for 10 s, 1% aq. Tween 20 for 1 min, 10% w/v calcium hypochlorite for 20 min. Proceed in aseptic conditions using a laminar flow hood.
2. Rinse the seeds three times in distilled water.
3. Put two seeds into distilled water in each of eight Petri dishes.
4. Germinate in the dark for 48 h.
5. Excise the apical 10 mm of the primary root and place into 20 ml of MR medium in a culture tube.
6. Incubate in the dark at 25°C. Lateral roots will develop after 5-6 days.
7. Remove and reculture lateral root tips.

Isolation and Culture of Roots of Monocots (e.g. Maize)

Equipment

Orbital shaker

24 × 90-mm sterile Petri dishes

2 × pairs forceps (Sterile) Filter paper (sterile)

Sterile culture tubes (20 ml).

Materials and reagents

White's basal salts—available commercially or make: H_3BO_3, 1.5 mg l^{-1}; $Ca(NO_3)_2.4H_2O$, 200 mg l^{-1}; ferric sulphate, 2.5 mg l^{-1}; $MgSO_4.7H_2O$, 360 mg l^{-1}; $MnSO_4.H_2O$, 5 mgl^{-1}; KCl, 65 mg l^{-1}; KI.0.75 mg l^{-1}; KNO_3, 80 mg l^{-1}; KH_2PO_4, 12 mg l^{-1}; $NaH_2PO_4.H_2O$, 16.5 mg l^{-1}; Na_2SO_4, 200 mg l^{-1}; sucrose, 2-4%; pH 5.8

3 × 250 ml conical flasks containing 200 ml of sterile distilled water

5 × 250 ml conical flasks containing 50 ml of White's medium (sterile)

24 maize kernels

250 ml conical flask containing 200 ml of 5% w/v sodium hypochlorite.

Procedure

1. Immerse kernels in 5% w/v sodium hypochlorite for 20 min Proceed in aseptic conditions using a laminar flow hood.
2. Rinse kernels 3 × in sterile distilled water (culture tubes).
3. Put two kernels onto sterile filter paper wetted with sterile distilled water in each of eight Petri dishes.

4. Germinate in the dark at room temperature.
5. Excise the apical 10 mm of the primary root and place into 50 ml of White's medium in a 250 ml conical flask.
6. Incubate in the dark at 25°C on an orbital platform at 50-60 rpm. Lateral roots will develop after 5-6 days.
7. Remove and reculture lateral root tips.

Hairy Root Cultures (Plant Transformation with *Agrobacterium rhizogenes*)

Equipment

Filter paper (Whatman no.1) 90 mm (sterile)
Scalpels (sterile)
Syringe and needles (sterile)
Forceps (sterile)
Conical flasks (250 ml, sterile)
Incubator for *A. rhizogenes* culture (28°C)
Incubator for Petri dishes - 22°C
Orbital incubator - 16°C dark, 120 rpm.

Materials and reagents

MS basal medium (sterile)

Petri dishes (90 mm) of 1/2 × MS basal medium + 3% sucrose and 0.2% Phytagel with Gamborg's B5 vitamins; 1 mg l^{-1} nicotinic acid, 1 mg l^{-1} pyridoxine HCl, 10 mg l^{-1} thiamine HCl, 100 mg l^{-1} myoinositol (sterile)

Petri dishes (90 mm) of 1/2 × MS basal medium + 3% sucrose and 0.2% Phytagel with Gamborg's B5 vitamins (sterile) and 10 μg ml^{-1} kanamycin and 250 μg ml^{-1} cefotaxime

Petri dishes containing yeast mannitol agar (YMA) medium: 2 mg l^{-1} $MgSO_4$, 0.5 mg l^{-1} K_2HPO_4, 0.1 mg l^{-1} NaCl, 10 mg l^{-1} mannitol, 0.4 mg l^{-1} yeast extract, 8 g l^{-1} agar (sterile) with 50 μg ml^{-1} rifampicin and 50 μg ml^{-1} neomycin added after autoclaving.

Antibiotic stocks (stored at - 20°C until used)
Rifampicin (50 mg ml^{-1} in methanol)
Neomycin (50 mg ml^{-1} in distilled water)
Cefotaxime (250 mg ml^{-1} in distilled water)
Kanamycin (10 mg ml^{-1} in distilled water).

Agrobacterium

Agrobacterium rhizogenes (e.g. strain LBA9402, Rlf®) with binary vector.

Procedure

Proceed in aseptic conditions using a laminar flow hood:

1. Surface-sterilize seeds and germinate in a sterile Petri dish.
2. Take an overnight culture of *A. rhizogenes* grown on YMA medium plates with 50 μg ml^{-1} rifampicin and 50 μg ml^{-1} neomycin at 28°C.
3. Take the seedlings and cut off the roots. Stab a sterile syringe needle first into the *A. rhizogenes* culture and then into the hypocotyl. Repeat for multiple inoculation sites.
4. Transfer the seedlings to Petri dishes containing 1 /2 MS medium with B5 vitamins.
5. Incubate for 48 h at 22°C, 16 h light.
6. Incubate at 22°C in the dark until roots form (7-10 days).
7. Transfer to 1/2 MS plates with B5 vitamins and 10 μg ml^{-1} kanamycin and 250 μg ml^{-1} cefotaxime.
8. The roots should be subcultured to fresh plates when required.
9. To establish a liquid culture, excise the apical 10 mm of the root and transfer to a 250 ml conical flask containing 50 ml of 1/2 MS medium with B5 vitamins. Incubate on an orbital incubator at 120 rpm at 16°C and subculture to fresh medium when the roots begin to become brown.

Sterile Culture of the Ear of a Cereal (e.g. Maize)

Equipment

Dissecting needles (sterile)

Scalpel (sterile)

Forceps (sterile)

Aluminium foil caps (sterile)

Incubator at 28°C.

Materials

(Sterile)

Dispensed into 125 ml conical flask (40 ml per flask): MS basal medium with White's vitamins and glycine (folic acid, 0.5 mg l^{-1}; nicotinic acid, 0.1 mg l^{-1}; thiamine HCl, 1.0 mg l^{-1}; pyridoxine HCl, 0.1 mg l^{-1}) glycine, 3 mg l^{-1}; l-inositol, 100 mg l^{-1}; $FeSO_4.7H_2O$, 27.8 mg l^{-1}, Na_2EDTA, 37.3 mg l^{-1}; sucrose, 6% w/v; kinetin, 10^{-7} M

3 × 250 ml beakers containing distilled water cpvered in foil

1 × 250 ml beaker containing 10% sodium hypochlorite.

Procedure

1. Take ears when 5-10 mm long.
2. Surface sterilize in 10% sodium hypochlorite (15 min).
3. Transfer to a laminar flow cabinet.
4. Rinse 3 × with distilled water.
5. Place ear into medium in conical flask.
6. Cover with sterile aluminium foil cap.
7. Incubate at 28°C, 18 h light, 6 h dark.

Embryo Rescue of a Dicot—*Phaseolus vulgaris*

Equipment

Mounted needles (2, sterile)
Pasteur pipettes (10, sterile)
Scalpel (sterile)
Dissecting microscope
Incubator at 26°C.

Materials and reagents

Ovule suspension medium Sucrose, 120 mg l^{-1}; agar, 0.75 g l^{-1} in distilled water

G1 medium in 90 mm plastic Petri dishes: Gamborg salts, NH_4NO_3 400 mg l^{-1}; sucrose 30 mg l^{-1}; thiamine HCl 1 mg l^{-1}; nicotinic acid 5 mg l^{-1}; pyridoxine HCl 0.5 mg l^{-1}; myo-inositol 100 mg l^{-1}; casein hydrolysate 1 mg l^{-1}; L-glutamine 1 mg l^{-1}; BAP 0.028 mg l^{-1}; Difco agar 8 g l^{-1}

R1 as 20 ml aliquots in 20 × 150 mm culture tubes: as G1, but without NH_4NO_3 and with L-glutamine 100 mg l^{-1}; casein hydrolysate 100 mg l^{-1}; GA_3 0.03 mg l^{-1}

Sterile distilled water (3 × 200 ml) in 250 ml conical flasks

Ethanol (200 ml 70% w/v) in a 250 ml conical flask

200 ml of calcium hypochlorite in a 250 ml conical flask.

Procedure

1. Harvest young seed pods 9-11 days after pollination.
2. Surface sterilize the pods in 70% ethanol for 1 min.
3. Immerse pods in 5% calcium hypochlorite for 5 min.
4. Rinse 3 × in sterile distilled water.
5. In a laminar flow cabinet, dissect out the ovules from the pods and place them in a drop of ovule suspension medium in a Petri dish on a dissecting microscope.

6. Isolate the embryos from the ovules using mounted needles and draw them gently in the ovule suspension medium into the tip of a Pasteur pipette.
7. Eject the embryo, with a drop of medium, onto the surface of the G1 medium in a Petri dish (up to eight embryos per dish).
8. Cover and incubate for 10 days in darkness at 26°C, then 20 days in 12 h light : 12 h dark at 26°C.
9. In sterile conditions, transfer the embryos to R1 medium in culture tubes.
10. Incubate 12 h light: 12 h dark at 26°C until seedling development is evident and a root system has developed. At this point, seedlings may be transferred to sterile pots containing 1:1:1 sand : vermiculite : compost and grown to maturity.

10

In Vitro Tissue Culture

The historic development of *in vitro* plant cell and tissue culture has undoubtedly been a major factor in the advancement of our knowledge of cell biology, physiology, biochemistry and more recently, molecular biology. However, its exploitation for applied purposes could be argued as being of even greater consequence. Plant biotechnology utilizes a range of *in vitro* techniques to manipulate plant germplasm, including clonal multiplication, generation of novel variants and the production of genetically modified plants through somatic hybridization and genetic transformation. In addition, *in vitro* culture can also have an important role in the conservation of plant germplasm. The use of *in vitro* germplasm storage in plant biotechnology programmes has a growing significance, as it improves the efficiency of research activities and secures the valuable products of such activities for both scientific and commercial purposes.

Conservation of plant germplasm can itself be the goal of *in vitro* plant cell and tissue culture programmes, by the use of techniques including micropropagation and embryo rescue. Tissue culture approaches have been vital in the re-establishment of endangered plant species, for example the lady's slipper orchid (*Cypripedium calcaeolus* L.). However, *in vitro* plant germplasm conservation requires an understanding and appreciation of the inherent problems of biotechnology and of the specific culture requirements of different plant species. In common with most *in vitro* plant cell and tissue manipulations, an overriding concern is the maintenance of the genetic fidelity of the stored germplasm. Thus, with time the phenotype and genotype of *in vitro* plant cultures change. Such changes constitute the basis of somaclonal variation, the significance of which to *in vitro* germplasm conservation is reviewed by Harding.

In vitro Propagation

Micropropagation is a general term which describes a variety of routes for the propagation of selected germplasm using *in vitro* techniques. Although the details of the different routes vary significantly in the mechanisms of plant multiplication, the five basic stages for successful micropropagation are comparable. The stages are:

- Stage 0. Preparative stage, involving germplasm selection.
- Stage 1: Establishment stage, involving the production of axenic, viable cultures.
- Stage 2: Multiplication stage, during which the number of propagules is increased.
- Stage 3: Plantlet production, involving the development of germplasm of sufficient size and quality for transfer to *in vivo* conditions.
- Stage 4: Establishment under *in vivo* conditions, involving the acclimatization of plantlets to glasshouse conditions.

Notable features of the *in vitro* propagation process are discussed in the following sections.

Germplasm Acquisition

The first step towards *in vitro* conservation is germplasm acquisition. For source plants in managed environments, collection can be reasonably straightforward. However, even for this germplasm it is important to consider factors, such as seasonal effects, which can significantly affect the ability to establish *in vitro* cultures. Before germplasm collection from either managed or non-managed habitats, the associated legal issues dealing with aspects of ownership, sovereignty and intellectual property rights have to be considered. It is also vital to ensure that target plants have been correctly identified and tissue samples can be collected from sufficient individuals to maximize the diversity of the collection, without endangering the remaining natural population.

In situations where germplasm is difficult to transport, or there is a significant risk of loss of viability during transit, *in vitro* collection provides a potential means of overcoming some of these difficulties. *In vitro* collection involves the placing of surface-sterilized tissue on to pre-prepared sterile culture medium in the field, prior to transport to the laboratory. To reduce the problem of *in vitro* culture

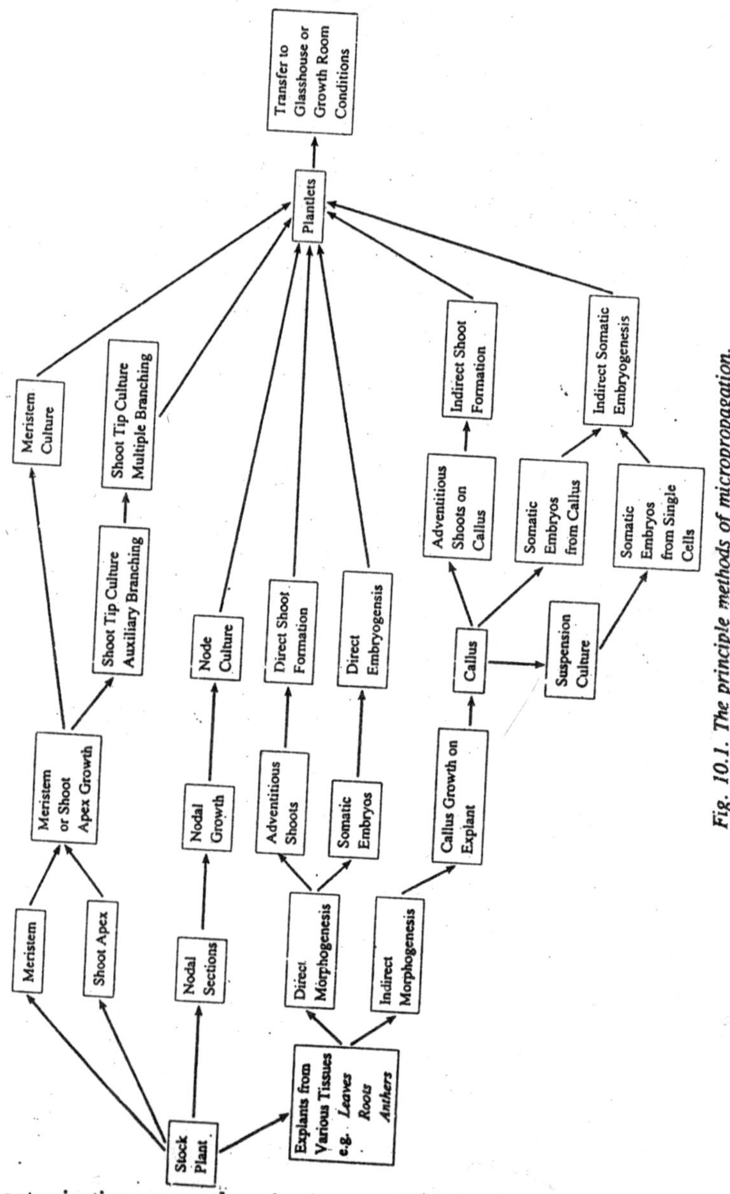

Fig. 10.1. The principle methods of micropropagation.

contamination, pre-prepared culture medium is often supplemented with antimicrobial agents. Techniques used in the field are kept as simple as possible to ensure that only rudimentary equipment needs to be transported. This approach has been successfully used for a number of species, including coconut, cotton and cacao.

Selection of Tissue for In vitro Culture

It has long been recognized that there are a number of factors which can significantly influence the *in vitro* behaviour of an explant, and which should be considered when selecting tissue for collection; these include:

1. The genotype of the source plant.
2. The explant tissue type.
3. The physiological age of the explant.
4. The season in which the explant was obtained.
5. Explant size.
6. The health and vigour of the source plant.

For some plant species a specific explant may be required for successful *in vitro plant* regeneration, for example immature and mature embryos are among the few explant types from which embryogenic cereal callus, a prerequisite for plant regeneration, can be initiated. The general consensus is that larger explants give better survival and regeneration rates as compared with smaller explants. However, the use of smaller explants does have the advantage of increasing the chance of virus elimination from subsequent cultures.

To reduce the significance of some of the above factors, appropriate pretreatment of the donor plant can be important. For example, the growth of rice anther and immature embryo cultures is influenced by the light regime the donor plants were grown under. In plants with strong apical dominance, removal of the terminal meristem can improve the *in vitro* response of lateral buds. Control of glasshouse or growth room conditions in which donor plants are maintained, for example by using irrigation systems which avoid soil splash, can reduce subsequent microbial contamination of *in vitro* cultures. Similarly, field-collected explants have been shown to carry more microbial contamination during the wet season as compared with the dry season.

In selecting explant tissue for *in vitro* culture initiation for conservation purposes it is important to consider the influence of different explant types on the occurrence and frequency of somaclonal variation in regenerants. In general terms the older and/or more specialized the explant the greater the potential for variation in derived plants. For example, plants regenerated from cultured petals of *Chrysanthemum* had higher frequencies of abnormalities compared with plants derived from pedicels. Such effects relate to changes in the genome, including endopolyploidy and DNA sequence amplification.

Microbial Contamination and Disease Indexing

Plant surfaces provide habitats for many types of microorganism. Additionally, plants may also have endophytic micro floras. Depending on the explant used, epiphytic and/or endophytic microorganisms can be carried over to *in vitro* culture. Surface microorganisms are normally removed by surface sterilization of the explant prior to culture initiation. The precise sterilization procedures are dependent on the explant type and plant species. Explants such as mature seeds and nodal stem pieces can be directly treated with sterilizing agents. However, tissue such as immature ovules and embryos tends to be easily damaged by disinfectants, therefore the ovary or ovule can be disinfected and the desired explants removed by dissection under sterile conditions. Although a range of chemicals have been successfully used to surface sterilize plant material most laboratories use diluted commercial sodium hypochlorite bleaches, such as Domestos and Clorox.. These preparations contain wetting agents which increase the effectiveness of the active compounds, they are easily available, cheap and relatively safe. After sterilization it is vital to remove the surface sterilizing agent by washing, normally with sterile water, to minimize toxic effects to the plant tissue.

Table 10.1. Examples and effectiveness of surface sterilizing agents

Disinfectant	*Concentration used (%)*	*Duration of treatment (min)*	*Effectiveness*
Calcium hypochlorite	9-10	10-45	Very good
Sodium hypochloride	0.9-2.0	10-45	Very good
Commercial bleach	5-30	10-45	Very good
Mercuric chloride	0.1-1	2-10	Good
Ethanol	70	2-3	Good
Antibiotics	4-150 mg l^{-1}	30-60	Quite good

Many of the endophytic microorganisms present within plant tissue will be capable of growing on plant culture medium, although some maybe inhibited by the high salt or sucrose concentration, or pH. Some types of endophytic microorganism will overgrow and kill slower growing plant tissues, while less adapted species can proliferate in the tissues of the explant utilizing nutrients from dead and damaged cells;

this can also lead to the death of an explant. Latent or subliminal microorganisms may overgrow plant tissue after transfer to fresh media, especially if the medium was reduced salt or sucrose concentrations. Even where plant cultures are not killed by overgrowth of subliminal microorganisms such contamination can affect the vigour of *in vitro* and *ex vitro* plants.

The ability to utilize *in vitro* cultures as a means of virus elimination is an important advantage of *in vitro* plant germplasm conservation. The technique of virus elimination pioneered by Morel and Martin (1952) was based on the principle that meristematic domes maybe free of viral particles. For example in sweet potato there is a direct correlation between explant size and elimination of the sweet potato yellow dwarf virus. However, not all viruses can be eliminated this way. Heat treatment (thermotherapy) in conjunction with meristem culture has been used to improve the success of virus elimination. The combination of meristem culture and thermotherapy has been used with a 97 per cent success rate in an *in vitro* cassava genebank at the Centro International de Agricultura Tropical (CIAT). Chemotherapy has also been assessed as a means of virus elimination, but only limited success has been reported, and the role of antiviral agents in virus suppression or inactivation has not been clear.

It is vital to define and maintain the phytosanitory status of *in vitro* cultures being used for germplasm conservation, especially those cultures that are to be involved in international exchange programmes. Therefore the availability of rapid, effective and sensitive procedures for disease indexing is vital. A number of approaches have been used, including enzyme-linked immunosorbent assay (ELISA) and nucleic acid hybridization.

Tissue Culture Media

Plant tissue culture medium usually consists of major inorganic salts, trace elements, a carbon source, vitamins, growth regulators and a gelling agent. The systematic study of the nutrient requirements of plant tissue under *in vitro* conditions has led to the development of a range of nutrient formulations which have now become the basis of commercial preparations used by most plant tissue culture laboratories. Optimization of *in vitro* growth is normally achieved by the modification of standard media formulae, which has resulted in numerous reports for a vast range of plant species. Significant factors in culture media formulation are discussed in the following subsections.

Table 10.2. Inorganic and organic compounds of four major plant culture media (mg l^{-1})

Compound	*Murashige & Skoog (MS)*	*Gomborg's B5*	*White's*	*Nitsch's*
$KHPO_4$	270			68
KNO_3	1900	2527.5	80	950
KCl			65	
Na_2SO_4			200	
$NaH_2PO_2 \cdot H_2O$		150	19	
$Na_2MoO_4 \cdot 2H_2O$	0.25	0.25		0.25
$NaEDTA \cdot 2H_2O$	37.3	37.3		37.3
$CaCl_2 \cdot 2H_2O$	440	150		
$CaCl_2$				166
$Ca(NO_3)_2 \cdot 4H_2O$			300	
$MgSO_4 \cdot 7H_2O$	370	246.5	750	185
$ZnSO_4 \cdot 7H_2O$	8.6	2.0	3.0	10
KI	0.83	0.75	0.75	
$NiCl_2 \cdot 6H_2O$			0.03	
$Fe(SO_4)_3$			2.5	
$FeSO_4 \cdot 7H_2O$	27.8	27.8		27.8
NH_4NO_2	1650			720
$(NH_4)_2SO_4$		134		
$CuSO_4 \cdot 5H_2O$	0.025	0.025	0.001	0.025
H_3BO_3	6.2	3.0	1.5	10
MoO_3			0.001	
$MnSO_4 \cdot 4H_2O$	22.3		5.0	25
$CoCl_2 \cdot 6H_2O$	0.025	0.025		
$MnSO_4 \cdot H_2O$		10.0		
Myo-inositol	100	100		100
Pyridoxine-HCl	0.5	1.0	0.01	0.5
Thiamine-HCl	0.1	10.0	0.01	0.5
Nicotinic acid	0.5	1.0	0.05	5.0
Glycine	2.0		3.0	2.0
Folic acid				0.5
Biotin				0.05
Sucrose	30,000	20,000	20,000	20,000

Sources and concentration of nitrogen

There are varied reported effects of the nitrogen composition of plant culture media. Explants from some species cannot tolerate high levels of nitrogen in the culture media, e.g. immature embryos of *Impatiens platypetala*. In rice anther culture medium the ratio of NH_4^+ and NO_3^- is critical for cell proliferation and plant regeneration, while the presence of organic nitrogen in culture media has been shown to enhance plant regeneration in, for example *Agrostis alba*. Hence most culture media contain a mixture of inorganic and organic nitrogen. The form and concentration of nitrogen in culture media have been shown to induce genetic variation in cultured plant cells. The number of albino plants regenerated from anther callus cultures of wheat is influenced by the potassium nitrate concentration in the culture medium. Furner et al. (1978) showed that the haploid *Datura innoxia* cell lines retained their ploidy in medium containing both inorganic and organic nitrogen, while cultures in medium containing only organic nitrogen were composed of diploid and tetraploid cells.

Carbon source

Sucrose, at concentrations of 2-5 per cent w/v, is the most commonly used carbon source in plant tissue and cell culture media. Higher concentrations have been utilized to induce embryogenesis and bulblet development in *Allium* sp. Increasing the sucrose concentration also provides a means of reducing tissue growth and has been the basis of several slow growth *in vitro* storage protocols for plant germplasm. Different types of sugar in culture media have been shown to enhance, for example plant regeneration and seed germination. Interestingly the maintenance of micropropagated plants under photoautotrophic conditions can result in the promotion of plantlet growth *in vitro* and *ex vitro*, while the resulting simplification of the micropropagation process can reduce labour costs.

Gelling agent

The most commonly used gelling agents for *in vitro* plant culture are agar, agarose and gellan gums, such as Gelrite. The physical and chemical properties of gelling agents have been shown to vary significantly between source and batch. As a result there are a growing number of reports indicating that media gelling agents influence culture characteristics, such as somatic embryo formation, shoot elongation, shoot multiplication, and hyperhydricity.

Plant growth factors

There are several groups of plant growth regulator substances, specifically, auxins, cytokinins, gibberellins, ethylene and abscisic acid. *In vitro* plant growth and morphogenesis are regulated by the interaction and balance between the growth regulators supplied in the medium, and those produced endogenously. Arguably the most important of the plant growth regulator groups are the auxins and cytokinins. Some typical responses include:

1. Promotion of auxiliary shoot formation by the presence of very low concentrations of auxins in combination with high concentration of cytokinins, for example in *Populus* sp.
2. Plant regeneration in monocotyledons is often promoted by transferring callus tissue to medium without auxins, or by replacing activity auxins such as 2,4-dichlorophenoxyacetic acid (2,4-D) with indole-3-acetic acid (IAA) or naphthalene acetic acid (NAA).
3. Promotion of adventitious shoot formation by the presence of low concentrations of auxins in combination with high concentrations of cytokinins, for example in *Rubus* species.

Although gibberellic acid (GA_3) tends to prevent the formation of organized root and shoot meristems in callus cultures, it can promote the further growth and development of pre-existing organs. For example, the presence of GA_3 in culture medium has been reported to improve the growth of potato meristems and inhibit callus proliferation, and increase the number of shoots produced from seedling and tuber shoot internodes of *Apios americana*.

However, it must be remembered that the effect of growth regulators can be significantly influenced by the basal culture medium. For example 0.5 mg l^{-1} 2,4-D in MS medium resulted in the production of approximately two plants from each embryo-derived callus of *Oryza sativa*, whereas on N6 medium with the same auxin concentration, up to nine plants per embryo-derived callus were produced.

Where *in vitro* plant culture is to be used for germplasm conservation it is also important to consider the role on the culture environment of the frequency of somaclonal variation. There is considerable evidence to indicate that somaclonal variation is influenced particularly by the type and concentration of plant growth regulators, and that growth regulators can act as mutagens. For example, 2,4-D has been shown to increase the frequency of blue to pink mutations in the *Tradescantia* stamen hair. Phenotypic variants in palms and African plantains have been attributed to the use of high concentrations of auxins and cytokinins respectively. 2,4-D has been shown to be more genetically 'damaging' than other auxins. For example, Jha and Roy (1982) showed that in suspension cultures of *Vigna sinensis* the maximum ploidy of cells grown in medium with NAA was hexaploid, compared with octaploid in cells in medium with 2,4-D. This phenomenon is particularly important in plants which are vegetatively propagated, apomictic and/or have long lived species, which would not normally or readily go through a sexual cycle which would eliminate 'epigenetic modifications'.

It is therefore vital to undertake a comprehensive literature search prior to starting *in vitro* culture to assist in the targeting of significant media components for study. Information from related plant species can often provide useful indicators for species for which *in vitro* culture has not been previously reported.

Problems of Culture Establishment

After excision from the donor plant, transportation to the laboratory, trimming and surface sterilization, the explant which is finally placed on the culture medium will inevitably contain stressed, damaged and dying cells. This results, particularly in many woody plant species, in the production of deleterious polyphenol and tannins, and lipid peroxidation products, which results in browning blackening of explant tissue and often cell death. Browning of tissues and the medium maybe prevented or reduced, by the addition, for example, of glutamine or activated charcoal to the culture medium, and/or preculture explant washing with antioxidant solutions.

Immediately after transfer to fresh culture medium, plant tissues release a range of substances, including alkaloids, amino acids, enzymes, growth factors and vitamins, into the culture medium. If the cells are inoculated at a low population density, the concentration of essential substances in the cells and in the medium can become inadequate for culture development, at which point the cultured cells can undergo programmed cell death. There is a minimum size of explant, or quantity of cells per unit volume of culture medium, for successful culture initiation. The size of the initial explant/cell inoculum also affects the initial rate of *in vitro* cell growth. The minimum inoculum density varies with plant species, explant type and cultural conditions.

Propagule Multiplication (Morphogenesis)

Once the explant tissue has been placed into culture, there are a series of different routes to the production of plantlets. The route which normally results in the most rapid multiplication of propagules is adventitious callogenesis. However, the involvement of a callus stage can result in a greater production of non-true-to-type plantlets as compared with propagation routes in which shoot regeneration does not involve callus. Hence this approach is the least suitable method for *in vitro* conservation. The use of shoot or meristem (shoot tip) cultures is preferred as the genetic fidelity of the germplasm is more likely to be maintained. However, not all shoots arising from shoot cultures originate from axillary buds. Adventitious shoots can frequently arise directly from shot tissue or indirectly from callus at the base of the shoot mass. Direct and indirect shoot development has been observed in apple shoot cultures. The occurrence of such shoots is increased when cytokinin concentrations are greater than the optimum concentration.

Indirect or direct morphogenesis occurs either by organogenesis or somatic embryogenesis. Although these are distinctly different processes, both depend on the ability to redirect cells and tissues which are mitotically quiescent, or already committed to some function or pathway of development, to a meristematic state, i.e. exhibit totipotency, that is the ability of an individual cell to regenerate into a whole organism. The development of adventitious shoot meristems is usually from the periphery of callus or explants, but cells from any histogen or cell layer can be involved. For many years these meristems were thought to have a multicellular origin. However, studies of plant regeneration from leaf discs of periclinal chimeras indicated that shoot organogenesis usually has a single cell origin. Two distinct types of somatic embryogenesis have been recognized. The first is direct embryogenesis, in which a single cell, or cell group commences meristematic activity and all the progeny of this cell form part of the embryo. This is a rare event and occurs, for example in citrus nucellular tissue. More common is indirect somatic embryogenesis in which somatic embryos originate from proembryonic cell masses, of single cell origin.

Whichever pattern of plant regeneration is being followed by an *in vitro* plant culture an event must initially occur that involves a change in the development of certain cells within the culture. Cells must acquire competence which allows the expression of organogenic or embryogenic potential. This change is referred to as an inductive event. The limited number of regeneration competent cells in an explant is illustrated by the reported difficulties of the production of transgenic grapevines. Grapevine leaves after co-cultivation with *Agrobacterium* expressed β-glucuronidase activity at the cut surface, in vascular bundles, or in inner cortical cells of the petiole, but none of these regions produce adventitious shoots. It is also important to remember that the ability of a culture to sustain morphologically competent cells declines with time in culture as a result of the effects of somaclonal variation.

The pattern of morphogenic development is determined by medium constituents and by genetic and epigenetic factors. Regeneration via standardized protocols can be restricted to specific cultivars; James et al. (1984) demonstrated that *Malus* rootstocks M25 and M27 regenerated by organogenesis from stem segments, but M9 and M26 did not regenerate. Genes from several plant species involved in plant regeneration *in vitro* have been identified in several plant species, including rice and orchard grass. Difficulty in regenerating certain genotypes could be a result of many possibly inter-related factors, including acquisition of competence, induction and differentiation. Each may be mediated in a different manner that requires separate investigation.

Plantlet Development

Basically there are only two options at this stage in the micropropagation sequence, to produce plantlets or cutting for transfer to the *in vitro* environment. For most plantlets, shoot elongation, whether combined with rooting or not, is still necessary and is usually achieved by transferring shoots or shoot clusters to cytokinin-free medium.

Acclimatization of *In vitro* Germplasm to *In vivo* Conditions

The quality of germplasm to be transferred to *in vivo* conditions is critical to the success of this procedure. Visual evaluation should be used to select out plantlets exhibiting disorders such as hyperhydricity or apex necrosis. The term vitrification was proposed by Debergh (1988) to encompass any disorder that could check the survival of *in vitro* plantlets once transferred to greenhouse or field conditions. It is common for substantial numbers of micropropagated plants not to survive the transfer from *in vitro* conditions. Losses of micropropagated plants are due to the lower humidities, higher light levels and non-sterile conditions of the *in vivo* environment. To ensure that plants survive and grow vigorously when transferred to compost an acclimatization process is required. This process normally requires the provision of a high relative humidity, by fogging, misting etc., shading, manipulation of the photoperiod and in some cases bottom heat. As the plants adapt to the *in vivo* environment the conditions can be gradually adjusted to the 'normal' levels. To assist rooting,

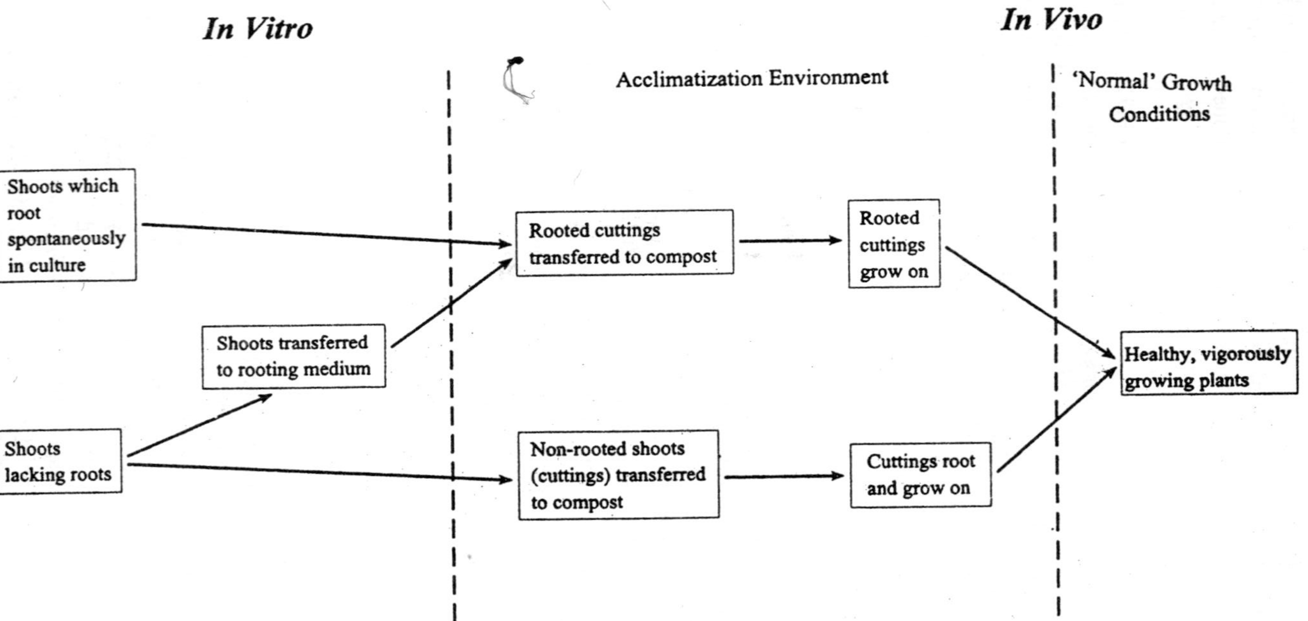

Fig. 10.2. Methods of rooting micropropagated shoots and in vivo establishment.

plantlets are often given an *in vitro* auxin treatment. The acclimatization process has been reviewed by Preece and Sutter (1991).

In vitro Culture Recalcitrance

Species of woody plants, cereals, grasses and some legumes are difficult to establish in culture, and recover viable regenerated shoots from, and are generally regarded as recalcitrant to tissue culture. Common problems include morphological recalcitrance, often in relation to tissue maturity and hyperhydricity. This has resulted in a range of empirical approaches in an attempt to overcome these difficulties, from the 'traditional', for example modification of growth factors types and concentrations in the culture medium, gelling agent, type of explant and the use of silver nitrate to control ethylene accumulation to more novel approaches, including the use of ventilated culture containers, bottom cooling and liquid raft culture. However, potentially of more significance have been the fundamental studies of the biochemistry and molecular biology of recalcitrant *in vitro* plant systems to try and understand the mechanisms of these problematic tissue culture responses. For example, cell ageing has been correlated with methylation of DNA sequences and demethylation implicated in the rejuvenation of woody perennials. The oxidative stress status of cultures appears to fundamentally affect *in vitro* plant development. It is from these studies that less empirical developments in plant tissue culture may be possible, with the aim of overcoming current recalcitrant responses. Such studies may also provide markers to tissue culture recalcitrance. However, the development of new approaches, which, if they are to be adopted by curators of *in vitro* genebanks, cannot be significantly more costly than existing procedures.

Embryo Rescue

Embryo culture or rescue involves the dissection of embryos from seeds and their 'germination' *in vitro* to provide one plant per explant. This technique has been used to overcome post-zygotic incompatibility which would otherwise hamper the production of desired hybrids. For example, Bodanesezanettini et al. (1996) used embryo rescue to recover hybrids between Brazilian soybean lines and wild perennial *Glycine* species. Embryo rescue can also provide a means of recovering seedlings from genotypes that have low and/or rapidly lost seed viability or protracted dormancy period and has proved to be of practical importance in the conservation of recalcitrant tree seed germplasm.

To control microorganisms, it is normally sufficient to surface sterilize the fruits or seeds, after which the embryo can be dissected out under aseptic conditions. To ease dissection, hard seeds can be soaked in water to soften them. The culture medium on which the excised embryo germinates is usually of a simple formulation, consisting of inorganic salts and sucrose. Use of immature embryos can result in the recovery of a greater number of seedlings as compared with mature embryos. However, they are more difficult to excise and have additional media ingredient requirements, such as amino acids, vitamins and growth regulators. Seedlings derived from embryo rescue are acclimatized to *in vivo* conditions in a similar manner to micropropagated germplasm. The method used in embryo rescue have been reviewed by Sharma et al. (1996).

Use of Plant Tissue Culture for Germplasm Storage

Reducing the growth rate of *in vitro* plant cultures can provide a convenient option for short- to medium-term germplasm storage. However, it is not suited to long-term programmes, because of risks of selection due to stresses imposed on the cultures during storage. A variety of approaches have been used separately and in combination to reduce the growth rates of *in vitro* plant tissue. Probably the most successful strategies have involved temperature reductions, but responses vary significantly both between and within species. For example, cold tolerant species such as strawberry and *Prunus* sp. have been successfully stored at 0°C to 4°C, but *Musa* plantlets cannot be stored below 15°C. *Coffea arabica* plantlets can be maintained at 27°C and only require sub-culture every 12 months, but *Coffea racemosa* plantlets have to be transferred every six months. A reduction in light intensity can be used in combination with temperature reductions, for example with banana cultures.

Modifications to the culture medium, including addition of osmotically active compounds such as mannitol, reduction of the growth factor concentrations, the reduction of the medium's nutrient status, and the use of growth retardants have all been reported to permit the maintenance of cultures in slow growth. Such changes to the culture medium have also been used in combination with reduced incubation temperature. The addition of activated charcoal to culture medium has also been reported as beneficial in minimal growth conditions. Roca et al. (1984) noted that its addition to cassava culture medium reduced defoliation, and limited chlorophyll degradation and root browning.

A reduction in growth of *in vitro* tissue can also be achieved by the lowering of the available oxygen levels. The simplest way of achieving this is to cover the culture with mineral oil. Overlaying callus cultures with mineral oil as a means of germplasm conservation was reported as early as 1959 by Caplin, who maintained 30 plant cell lines under mineral oil, with subculturing every 3-5 months, for longer than three years without apparent change in growth characteristics after transfer to normal growth conditions. It has been noted that mineral oil overlay could provide a useful means of short-term plant germplasm conservation in low-tech situations.

Slow growth is used as a short- to medium-term storage approach in many laboratories, international/regional centres, including Centro Internacional de la Papa (CIP) and CIAT. However, even with increased time between subcultures, management of large *in vitro* collections is problematic. There is also continued concern about the level of somaclonal variation under slow growth conditions. However, cassava stored under slow growth conditions for 10 years has been shown to remain genetically stable.

Recovery of Germplasm after Storage

Cryopreservation of plant tissue is the only viable option for the long-term storage of germplasm from species which are vegetatively propagated or produce recalcitrant seeds. Cryopreservation will be discussed at length in other chapters in this volume, but it is important to remember that cryopreservation depends for its success to a significant degree on appropriate tissue culture approaches. Germplasm to be cryopreserved should be healthy and disease-free, therefore an understanding and appreciation of the *in vitro* requirements of germplasm to be cryopreserved is important. On thawing, plant tissue inevitably suffers damage and stress. The composition of culture medium has been shown to significantly affect post-thaw cellular cryoinjury, for example lipid peroxidation. Therefore the use of specific post-thaw culture media can be important in the initiation of cell regrowth. For example, the use of ammonium ion-free medium and supplementing the culture medium with activated charcoal and the iron chelating agent desferrioxamine have all been shown to improve post-thaw cell regrowth.

Facilities for Plant Tissue Culture

There are many descriptions in the literature of the requirements of laboratory and facility design, such as Mageau (1991), Bhojwani and Razdan (1996), which can be extremely elaborate and expensive both in terms of capital expenditure and running costs and can be unnecessary for the tissue culture being undertaken. At the early planning stages it is important to ensure that the basic infrastructure, such as continuity of power and supply of appropriate quality water are in place, and that local back-up support services are available for the repair and servicing of equipment. Key requirements of any plant tissue culture facility are:

1. An area for media preparation, washing up and sterilization, equipped with standard laboratory equipment for preparation of solutions such as pH meters and stirrers, an autoclave and glassware washing and storage facilities.
2. A transfer area for manipulation of the *in vitro* culture, in which prepared sterile media and equipment can also be stored. This room should be equipped if possible with sterile transfer hoods or UV light boxes for manipulating *in vitro* cultures.
3. A growth room with an air-conditioning unit and open shelves, to aid air flow, preferably of metal construction with fluorescent tubes running under the shelves. It is preferable that the growth room is situated near the transfer area, in a room without windows and if possible away from outside walls to reduce the effects of external temperature fluctuations.

It is vital that the design of the transfer area and growth room is such that both areas can be easily kept clean. This is an important consideration in avoiding loss of cultures due to microbial contamination.

CONCLUSIONS

Tissue culture provides a potentially important means in itself for the *in vitro* conservation of plant germplasm and also vital supportive technology for long-term cryopreservation-based storage. However, its success requires understanding and an appreciation of the shortcomings of plant tissue culture, particularly in terms of minimizing the effect of somaclonal variation. No one tissue culture approach is universally appropriate for the storage of all plant germplasm. Therefore the development of integrated approaches will be important, such as that discussed by Dixon (1994). Finally the cost implications of *in vitro* approaches to plant germplasm conservation have to be compared with more traditional approaches. In a recent cost analysis study of the field and *in vitro* parts of the cassava collection at CIAT, it was shown that the *in vitro* collection, including the isozyme laboratory for genotype characterization was 53 per cent more expensive than field-based collections. However, the authors felt that the annual cost of preserving the world's most complete collection of cassava germplasm was not excessive considering the importance of the crop as a major food source.

11

In Situ Subculture

Investigations of intracellular changes have been conducted on cells released from contact inhibition by reseeding cells in a larger growth area after trypsinization, by addition of fresh serum, or by transforming these cells. These methods of releasing contact inhibition pose some disadvantages for the interpretation of experimental results. It is uncertain, for example, that early biochemical changes seen after reseeding trypsinized cells are due solely to loss of contact inhibition. Some may represent changes necessary to repair cellular damage caused by the trypsinization.

Serum stimulation in contact-inhibited cells does not cause physical cell separation, and biochemical alterations seen after addition of fresh serum to a confluent culture might be due to stimuli other than those operating in an authentic release of contact inhibition. Similarly, virus infection of contact-inhibited cells can and does activate processes other than those simply concerned with cell to cell contact.

A method of releasing contact inhibition without disturbing cell substrate attachment, or the chemical composition of the medium, should provide for an increase in the space available for growth of a confluent culture. Ideally, such an increase in growth area should be uniformly available to every cell within the culture. Increasing the growth area along a border of the culture leads to stimulation of cell division in only a very small and insignificant fraction of cells immediately adjacent to the border. Cells within the center of the culture are not released from contact inhibition. In this laboratory we have studied the macromolecular changes accompanying release of contact inhibition by using a method by means of which a large fraction of cells in a confluent monolayer are released from contact inhibition of growth and division. This is accomplished without chemical treatment of the cells, without change of medium, and in such a way that most cells are provided with free growth area. The procedure involves growing the cells to confluence on surfaces uniformly covered with glass beads 200 mm in diameter. When confluence has been attained, contact inhibition may be released by discarding the beads, leaving behind numerous spaces throughout the culture. Removal of the beads dislodges few if any of the cells. After release of contact inhibition by removing the beads, the cultures double in cell number following the first round of DNA replications, and continue to grow until they are again contact inhibited. Cell types used have included several cell lines and strains including both primary cell cultures of neonatal rat heart cells and established 3T3 and 3T6 mouse fibroblast cell lines.

Cell Cultures

Cells are cultured in Dulbecco's medium with the addition of 10% calf serum, 75 units/ml penicillin, and 50 mg/ml of streptomycin. The cultures are incubated at 37°C in a humid atmosphere of 5% CO_2 in air. Cultures are checked periodically for contamination by microorganisms.

Cultures are seeded in the presence of glass beads. One gram of glass beads (200 mm average diam., 3M), is placed in a 60-mm plastic tissue culture Petri dish after dry heat sterilization of the beads for 6 to 12 hours at 170°C, The cells in 5 ml of medium are then seeded into the culture dish with the glass beads. By gentle tipping and swirling motions an even layer of beads is made to cover the bottom of the dish. The dish is carefully placed in the incubator and after a short time the cells settle between the beads. The cells attach on the surface of the dish between the beads and multiply until the spaces between the beads are completely occupied by adherent contact-inhibited cells.

Medium in the Petri dish is changed every 3 days. The spent medium is removed with a curved-tip Pasteur pipette fitted with a rubber bulb, slowly and carefully so as not to disturb the beads. Only about 85% of the spent medium is removed. If the dish is drained dry, some beads float off the surface when fresh medium is returned to the dish. The fresh medium is added slowly to the dish through a curved-tip pipette lying against the side of the dish.

Removal of Beads

When the surface of the Petri dish between the glass beads is fully covered with cells, and at least 48 hours after a change of medium, contact inhibition of cell division may be released by discarding the glass beads. The culture dish is tipped slightly to one side. The medium is sucked up into a wide-bore pipette and then allowed to drip back onto the dish so as to wash all the glass beads to the low side of the tipped dish. The beads and medium are then pipetted together into a centrifuge tube which is spun at 1600 g for 1 minute to pellet detached cells, dead cells, and large debris with the beads. The cell-free supernatant medium is then returned to the tissue culture dish, which now contains a cell layer interspersed throughout with holes left by the discarded beads.

Release of Contact Inhibition

When the beads are discarded, the cells in the dish start to multiply and fill up the holes. Within 12 to 36 hours, depending on the cell type, mitotic cells in large numbers can be seen within and at the periphery of the holes. In 72 to 96 hours a confluent monolayer of contact-inhibited cells has been formed. A plot of DNA synthesis (thymidine incorporation) and mitotic index in the hours following release of contact inhibition by the glass bead method. There is a semisynchronous burst of DNA synthesis with a peak at about 30 hours after release followed by a semisynchronous peak in the mitotic index at about 40 hours.

Cell Growth

Normal cell growth characteristics are preserved in the presence of the beads and after their removal. Media are changed typically 2 days after inoculating the cultures and every 3 days thereafter. The cell cultures grow logarithmically until confluent at about 10^5 cells per plate and remain contact inhibited until the beads are removed. After removal of the beads, the cells once again multiply logarithmically as shown in the figure. No change of media at times prior to removal of the beads is necessary for release of contact inhibition since growth curves similar to that seen here can be obtained without any change of medium throughout the period of growth with beads, removal of the beads, and reestablishment of a confluent culture at a higher cell density.

Comments

Certain considerations and precautions should influence use of the glass bead technique to maintain and release contact inhibition of cell division. The beads must be sterilized with dry heat for at least 6 hours at 170°C. Sterilization in steam leads to caking of the beads so that they cannot be dispersed as a uniform layer. Sterilization for less than 6 hours occasionally leads to growth of resistant microorganisms.

The use of beads smaller or larger than 200 mm has given less desirable behaviour of the contact released monolayers. If the beads are smaller than 200 mm they do not settle well in the culture dish and do not provide sufficient space for growth between the beads. If the beads are larger than 200 mm there are many more cells between beads and a lower percentage is therefore released from contact inhibition when the beads are removed.

The presence of glass beads in intimate association with the cells has no apparent effect on their characteristic morphology, as determined by phase microscopy before and after removal of the beads. The glass beads used are composed of a soda-lime glass which does not provide a good substrate for growth of tissue culture cells. Therefore cells grown in the presence of the beads do not adhere to the beads and are apparently not damaged physically to any significant extent when the beads are removed.

Normally, beads are discarded at least 48 hours after medium change and the same medium returned to the cultures to exclude the effect of releasing contact inhibition by medium change and fresh serum addition.

Mitotic indices, in our experience, are best obtained from mitotic cell counts using X 150 phase microscopy. The use of fixatives and stains in the counting of mitotic cells leads to an underestimation of mitotic cells. Mitotic cells round up and show decreased adhesiveness to the plastic substrate upon which they are growing. Agitation caused by adding fixatives and stains dislodges a great many of these. Observation of dislodged cells show that greater than 50% are in mitosis.

12

MICROPROPAGATION

The raising of young plants is of vital significance for whatever purpose a crop is being grown. Natural ecosystems regenerate themselves gradually during a period of time, but almost all agriculture and horticulture relies at some point on the establishment of fresh stocks of plants. It must be remembered that most of our food comes from annuals which are propagated by seed, i.e. through sexual reproduction.

Table 12.1. The percentage contribution to World food production of the major crops

Crop	%	*Crop*	%
Wheat	15.7	Soya bean	2.3
Rice	13.0	Sorghum	2.0
Maize	12.5	Leguminous grains	1.9
Potatoes	10.8	Millet	1.9
Barley	7.1	Oats	1.8
Sweet potatoes	5.1	Tomatoes	1.5
Cassava	3.9	Rye	1.0

On this list only potatoes and sweet potatoes are normally propagated by means other than seeds. Food crops for which vegetative propagation is common tend to be horticultural ones grown on a relatively small scale (e.g. fruit trees).

Vegetative propagation is expensive. Potatoes are exceptional since they naturally produce large numbers of tubers which can be used to give rise to the following year's crops, but even here the transport costs for seed potatoes to plant one hectare are much higher that for a wheat seed to sow a one hectare crop. For species which require the taking of cuttings, labour costs can be considerable. Vegetative propagation is thus only employed for high value crops where its advantages are particularly worthwhile.

Vegetative propagation results in genetic uniformity, i.e. clones of plants. This means that once selections for desirable characteristics have been made, all the new plants raised will be identical. If the seeds of many fruits are sown, frequently the resulting plants will be quite unlike the plants from which the fruit came. The plant stock for fruit orchards is normally vegetatively propagated to avoid this problem (unless breeding work is being undertaken in the orchard to develop new varieties).

Ornamental horticulture also relies heavily on vegetative propagation since flower colour and form (e.g. in lilies) are frequently susceptible to change following sexual reproduction.

It must, however, be remembered that after several generations of self-fertilization (in plants where this is possible) an almost genetically uniform population will be produced; seeds from such plants may be almost as similar to their parents as plants propagated vegetatively.

Man has cloned plants by vegetative propagation for thousands of years and a wide range of techniques, employing various plant parts, has been developed for different species. The application of cell culture techniques *in vitro* means that the traditional taking of cuttings etc may be extended to new fields. In relation to micropropagation, the specific advantages of culture *in vitro* are summarized again here:

1. The rapid propagation of species difficult to multiply by conventional vegetative means. This permits the production of elite clones of selected plants;
2. The propagation of newly bred cultivars often of ornamental plants. These techniques enable large numbers of plants to be brought to the market place quickly, enabling a quicker return to be made on the investment that went into the breeding work;
3. The production of disease-free (particularly virus-free) plants. For low unit value crops, such as potatoes, the initial tissue cultured stock may then be further propagated by cheaper conventional means (with increased risk of virus re-infection);
4. In plant breeding experiments, selected lines may be created by haploid and protoplast work. These may then be propagated indefinitely *in vitro*;

5. The propagation of plants rescued from environmentally threatened areas; only small amounts of tissue need to be removed from the wild.

Note that propagation *in vitro* may be carried out at any time of the year once clones are established; conventional techniques are often season dependent.

Such propagation tasks are the province of the micropropagation industry. This originated in the 1970s and 80s with the establishment worldwide of a large number of small firms, mainly linked to established nursery businesses. As a result of experience and competition these have now been reduced to a smaller number of larger concerns. While their long-term viability is often tenuous, currently micropropagation represents the only area of plant biotechnology which has found widespread commercial application.

Strategies for Propagation *in vitro*

Toshio Murashige was one of the pioneers of plant cell culture *in vitro*; and was largely responsible for the development of the widely used MS medium, Murashige proposed that micropropagation could be divided into a number of distinct stages.

Table 12.2. Stages of micropropagation

Stage 0	Selection and preparation of mother (stock) plants—sometimes called donor plants
Stage I	Establishment of the aseptic culture
Stage II	Multiplication of propagules
Stage III	preparation for re-establishment of plants in soil (rooting)
Stage IV	Transfer to natural environment—i.e. *ex vitro* (weaning)

Many of the basic techniques necessary to accomplish each of these stages they will be given only brief consideration here. The designation of Stage 0 emphasizes the importance of healthy stock plants for the success of the subsequent cultures. The growth of these plants in conditions that are as clean as possible greatly aids the establishment of aseptic culture.

The overriding consideration is to use plants which have been raised in an environment in which contamination can be kept to a minimum. Plants from glasshouses are preferable to those grown outdoors, particularly if they are watered from below, to avoid soil being splashed onto the leaves. Effective control of pests and diseases permits vigorous explants to be taken which are less likely to harbour saprophytic micro-organisms. For species where this is possible, shoots may be forced (in the laboratory) from dry rhizomes to avoid all contact with soil.

For a given species, the selection of plants in a particular phase of growth may have an effect on the success of culture establishment. For example, many more shoots are produced from juvenile ivy explants than from mature ones. Nutrition is also important—exceedingly vigorous tomato plants grown with abundant nitrogen fertilizer maybe less prolific shoot producers in culture than more poorly fertilized specimens. Since light intensity and photoperiod are known to have many effects on morphogenesis, it is not surprising that they are of importance here. For example *Azalea* microcuttings root best when derived from stock (or donor) plants grown in dim light. Similarly, the temperature regimes to which plants are exposed may be critical. This could sometimes be due to thermal shock stimulating differentiation in the explanted tissues. It is important to remember that the way the stock plants are handled can have profound effects on the success of the cultures later established from them. Published protocols frequently do not describe the history of such plants in detail. This may account for the difficulties often encountered in trying to repeat them.

Although reports of the use of plant growth regulators in culture medium abound, their administration to stock plants is more seldom reported. It may, however, give beneficial results and can have the advantage that the time in aseptic culture is reduced. Spraying stock plants of tomatoes with the antigiberellin chlormequat chloride (CCC) can enhance subsequent shoot proliferation.

Dipping leaves of petunias into benzyladenine (BAP) solutions for 30 seconds before cutting into explants has a similar effect. Pruning treatments can also enhance shoot production, presumably because of modifications to endogenous plant growth regulator concentrations.

The multiplication stage is the heart of any micropropagation system. Repeated cycles of subculturing may continue indefinitely. The proliferation rate determines the economic feasibility of propagating a given species. Most commercial practices currently employ either axillary branching or adventitious shooting (or a mixture of both) for the majority of species. Steps involving callus formation have generally been avoided. In spite of the many published protocols for propagation by somatic embryogenesis, these have been rarely used commercially.

They do, however, often feature in plans for automated production systems.

Micropropagated plantlets can only be transferred to the non-axenic environment with care. They must become adapted to lower relative humidity and possibly be permitted to undergo a dormant stage. With some plants, such as potatoes, a special rooting stage is not required since adventitious roots develop during multiplication or in the compost without extra stimulation. The plantlets may be treated like cuttings, excised and dipped in an auxin-containing rooting powder. For many woody species the rooting stage is one of the most difficult stages to accomplish; several medium changes may be necessary, since auxins stimulate root initiation but inhibit root extension. Some phenolic substances have been shown to have beneficial effects in certain situations (e.g. phloroglucinol with apples). The physiological state of the tissues affects their ability to root. This state can be modified by the subculturing regime adopted: in general, the juvenility associated with repeated subculturing makes root induction easier.

The stages described above permit the process of micropropagation to be broken down and the problems with different aspects to be considered separately. It must be remembered that in a commercial situation there must be a smooth flow through the propagation process from beginning to end. This is especially so since plantlets are delicate and cannot be stockpiled part way though the process as is possible with the products of engineering!

VARIATION

One of the chief aims of micropropagation is to produce uniform plants, yet the danger of plant-to-plant variation is one of the commonest criticisms levelled at it.

In theory, vegetative propagation of any type should result in the production of totally uniform plants, since no rearrangement of the chromosomes by meiosis is involved. However, should a mutation (a spontaneous change in the DNA sequence) occur, cell division will result in a clone of cells of the new type appearing in the population. (This can happen in the buds of intact plants in nature to produce what are sometimes called '*sports*'). The result of this is somaclonal variation (soma—refers to the cells of the body of an organism other than the gametes or gamete-producing cells). We remind you that regenerated plants differing from the starting material should properly be described as variants; the term mutant should not be applied until sexual crosses have established that the changes are truly due to genetic

(DNA) change. This is because plants can be phenotypically altered (epigenetically changed) as a result of the culture procedure. The elimination of some viruses can also adversely affect the regenerated plants (e.g. strawberries may produce more runners but smaller fruits).

Because of the nature of cell culture *in vitro* the occurrence of somaclonal variation may be more frequent than expected. Here we shall only consider its accidental occurrence arising during micropropagation. This somaclonal variation is influenced by a number of factors:

1. *Genotype* different cultivars of one species may be more prone or less prone to somaclonal variation than others;
2. *Explant source* in geraniums for example, variants have been recovered from roots and petals but not from stem cuttings. Many plants normally propagated vegetatively (e.g. sugar cane and potato) are not genetically uniform but are chimaeras (genetic mosaics). Obviously, regeneration from such plants can result in rearrangement or loss of some of their cell lines. For this reason ornamental variegated plants are particularly difficult to propagate by tissue culture;
3. *Age of cell culture* it is widely accepted that most long-term cultures are chromosomally variable, particularly cell suspension and callus cultures. Regeneration occurs mainly from the normal cells, but polyploid (with multiple chromosome sets) or aneuploid (with incomplete multiple chromosome sets) plants can be produced;
4. *Culture conditions* the growth regulator 2,4-D used in the propagation medium has been most widely implicated in inducing genetic change in many crops.

2,4-D is known to be a *mammalian carcinogen*. In other words it appears to encourage the insertion of incorrect bases into DNA (mutations). Other synthetic auxins and high concentrations of the cytokinin BAP have also been found to increase the number of variants regenerated. It may be seen that variation in micropropagated plants is due both to the changes induced by the culture regime itself and to the expression of pre-existing variation in the explant. This must be borne in mind when deciding on the procedures to adopt in order to produce uniform propagules.

Transfer to Non-axenic Conditions

Shoots grown *in vitro* with or without roots are not adapted physiologically to a life in soil and, so must be gently weaned to accept such conditions.

The two major dangers are that the plantlets may become desiccated and may be subjected to fungal rot. The reasons for this can be explained in the following way. The relative humidity in culture containers is usually close to 100%. As a result of this the plantlets do not need to regulate their transpiration rates. Consequently their stomata remain open all the time and they only regain the ability to close these stomata over a rather extended period. At the same time the layer of epicuticular wax is thin and the intercellular leaf spaces are large. In addition the roots formed *de novo* are often only poorly connected to the plant's vascular system and root hairs are rarely present. Transfer of such plantlets to soil in a relatively low humidity atmosphere results in desiccation. To avoid this, high humidity is maintained by fine jets of water (*misting* or *fogging*), but this can encourage the other major problem—fungal rots (damping off). The compost must be sterilized and all sucrose containing agar medium washed off the plant roots before transfer to soil. A gradual reduction in the relative humidity around the plants during a period of several weeks is normally sufficient, but in some cases desiccation problems may be so severe that the leaves grown *in vitro* must be cut off to prevent transpiration. This has the obvious danger that the small plantlets will not be left with sufficient reserves to regrow new leaves. A possible compromise would be to spray them with a chemical film that would block up the stomata. An alternative approach is to reduce gradually the relative humidity *in vitro* during the rooting stage. Porous lids may be used to cover the vessels to allow a slow acclimatization, although to prevent the medium from drying up it must be replenished frequently with sterile water.

Normally under-soil heating cables are used to promote good root growth and careful attention must be paid to the normal control of insect pests and diseases.

Alternatives to this method of propagating microcuttings have been investigated. These mainly involve the use of somatic embryos to create synthetic seeds—also called *somatic seeds* or *pseudo seeds* or *artificial seeds*. The aim here is to replace the endosperm and seed coat of true seeds with a protective and nutritive gel which can encapsulate the embryo. Alginate gels are particularly attractive for such work because they can form so simply but there are also other possible encapsulation agents such as the polymer known as *Polyox*.

This technique has been extensively investigated with crops such as alfalfa, celery, cauliflower and carrot, but to be successful the embryos must be at the right stage of growth. Encapsulation has given

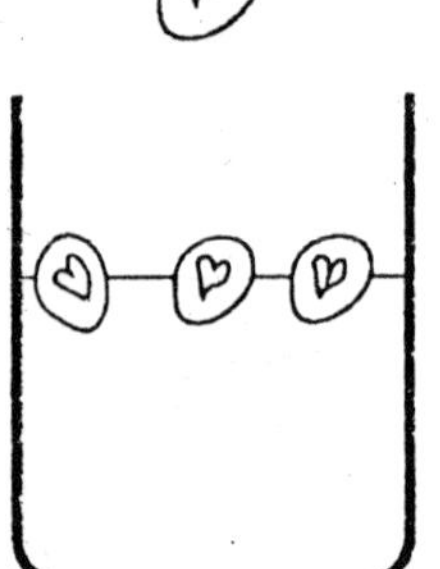

Fig. 12.1. The scheme for encapsulation of somatic embryos.

good results with buds of mulberry produced *in vitro*. It is possible that the beads produced in such a manner could be used in existing fluid drilling machinery for pregerminated seed; this would considerably reduce the costs of delivery of micropropagated crops to the field.

One of the advantages of true seeds is that they usually become dormant between maturity and germination. Providing they are kept dry, most species can be stored for a relatively long period until they are required. Much interest has surrounded the induction of dormancy in somatic embryos so that their production may be independent from field planting. To some extent this has been possible (by desiccation and growth regulator treatments) in carrot and grape. Much more research work, however, is required before this somatic seed technology will become available on a commercial scale.

Problem of Labour

The multiplication stage for most commercially micropropagated plants involves the manual cutting apart of shoot clumps and placing them on fresh, usually agar-gelled, medium. This is clearly time-consuming work. For asepsis to be maintained considerable manual care is required and labour typically represents up to 60% of the total production costs. If this figure could be reduced it would become economically advantageous. It would also lead to greater commercial viability for a wider range of plants to be propagated *in vitro*. One approach to cost reduction is just to relocate the micropropagation unit to an area of cheap labour. This, however, may be unpopular politically. An alternative strategy for cost reduction involves mechanization and automation. A number of companies are now marketing systems for automatic medium preparation and dispension although these are not the most expensive parts of conventional technologies. Similarly, automated surface sterilization of explants, controlled by computer, has been reported but this operation also represents only a small part of the running costs of a commercial micropropagation establishment.

Sub-culturing of explants by machine would be commercially most rewarding, but is much more difficult to achieve.

It is possible that contamination problems would be reduced since, once initiated, the cultures need not be exposed again to the glasshouse environment until weaning, Conversely if contamination is accidentally introduced into such a system it could rapidly spread with devastating results, rather than being isolated to a few culture containers as happens with manual systems.

Most ideas concerning automation have centred on the use of liquid media. Agar gelled medium is most commonly used in conventional micropropagation but a system in which the medium can be pumped away and exchanged for a new one lends itself more to a reduction in labour costs. Obviously, a huge range of designs of containers and machinery is possible but it is important that adequate attention is paid to the biological side of culture. Our knowledge of how to control the differentiation of the tissues is rather inadequate; so many proposed schemes are pieces of clever engineering, but fail because the plants do not develop reliably enough for commercial success. Simplicity is often advantageous.

Automation is largely a feature of the future. In the following discussion we will describe some of the approaches that have been tried and speculate on some future possibilities. A first step towards automation has involved the use of shaken cultures in liquid medium, often in conical flasks. Many plants (e.g. pineapple) show enhanced

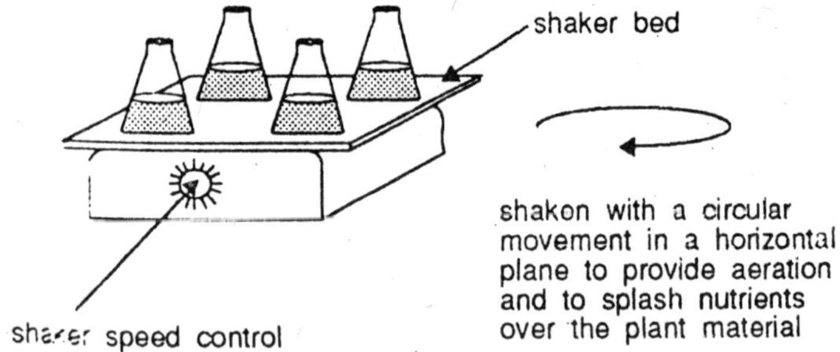

Fig. 12.2. Orbitally shaken liquid shoot cultures.

rates of shoot proliferation under such conditions, but the shaker beds required take up much more room than ordinary shelving used for agar-based media.

The term *bioreactor* means different things to different people, but may be considered as an apparatus to provide all the needs of a growing culture. Some bioreactor designs support the explants upon an inert surface (such as a plastic mesh) and circulate the medium around them. There are a number of possible refinements. The medium may be allowed to circulate continuously or be replenished at intervals

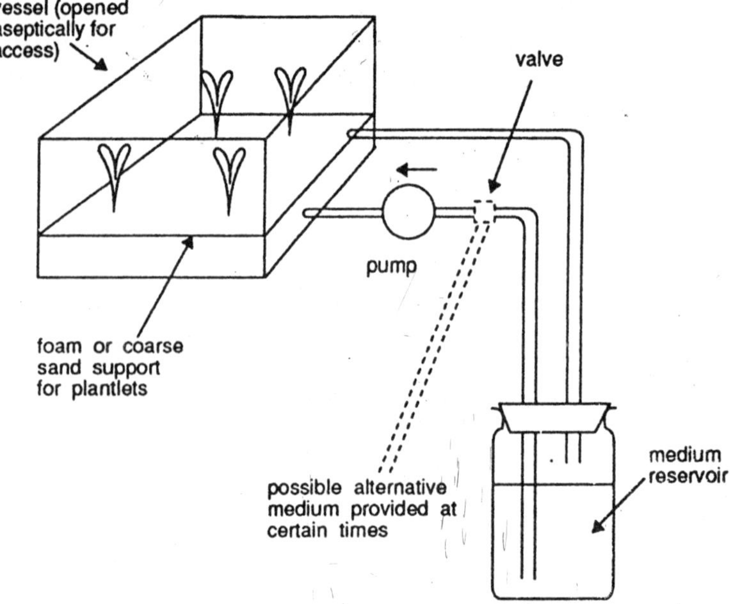

Fig. 12.3. Bioreactor scheme containing immobilized shoot explants.

possibly automatically, in response to the depletion of a particular nutrient or the accumulation of an easily detectable biochemical waste product, or to a significant change in the pH of the medium.

The culture abnormality known as *vitrification* was described earlier; amongst other factors the high humidity associated with liquid media is known to encourage its development. While vitrification is deleterious in that the plantlets are not suitable for weaning, there is some evidence that in the multiplication stage it can result in increased productivity. However, it must be reversed before the rooting stage. Excellent growth of many types of tissue has been reported when the medium is supplied as a fine aerosol and allowed to envelop the explants as a fog or mist. Again, the medium is recirculated and it is easy to change its composition in order to supply a sequence of treatments. A disadvantage often encountered with this type of bioreactor is the narrow nozzle of the sprayer which may easily become blocked.

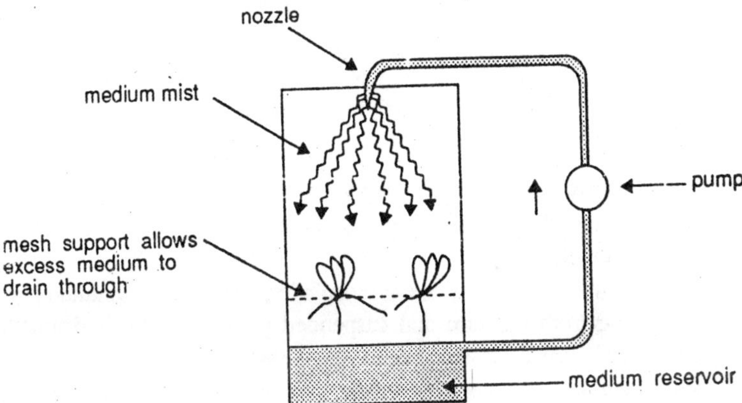

Fig. 12.4. Bioreactor scheme for supplying medium as a mist.

Some tissues (e.g. gladiolus bulblets) have been successfully grown without support in submerged liquid culture, agitation either being provided by aeration alone or by additional stirring. Many workers use apparatus designed and marketed for microbial fermentation because so little is made specifically for plant work. This may be satisfactory, but frequently greater success (for much less cost!) results from improvised designs made especially to cater for the growth of plant tissues. There are three main ways:

1. *Batch culture*—here the vessel is inoculated with plants which are then allowed to grow. After a period of time they are all removed

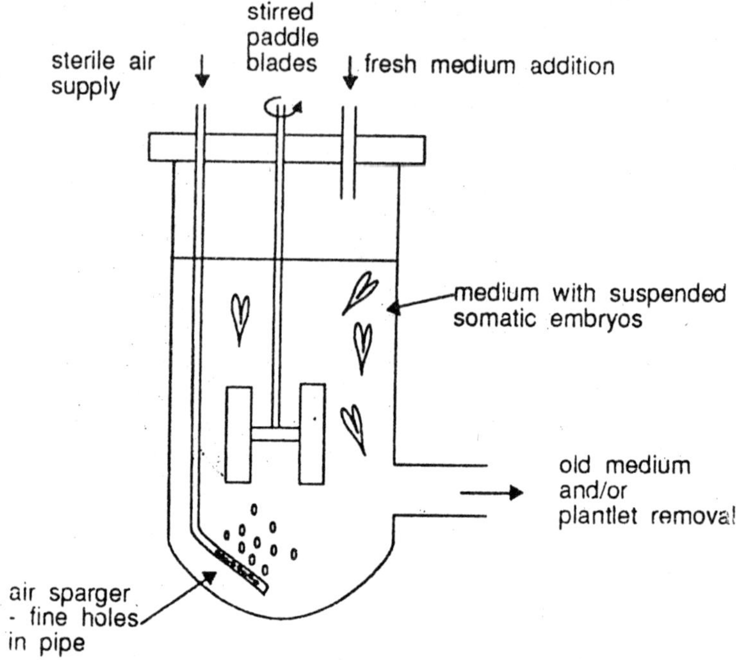

Fig. 12.5. Stirred tank bioreactor scheme to provide submerged liquid culture.

together and the vessel is re-autoclaved and made ready for a fresh inoculation.

2. *Continuous culture*—after initial inoculation, medium is continuously added and both medium and suspended plant material is drained off to maintain a steady state in the vessel.
3. *Semi-continuous culture*—movement into and out of the vessel takes place periodically. At intervals of up to one month, up to 80% of the culture is removed and replaced with fresh medium thus now the apparatus always contains some living material.

Great design flexibility is possible. Plant material may be size segregated and medium separated by straining through meshes of various sizes. Several vessels may be linked together and various types of instrumentation linked to a computer to control the environment within the reactor system.

Somatic embryogenesis usually requires a complex series of different media, particularly if the embryos are to be desiccated and used to create somatic seeds by encapsulation. Since they are discrete structures they also lend themselves to being pumped gently from one vessel to

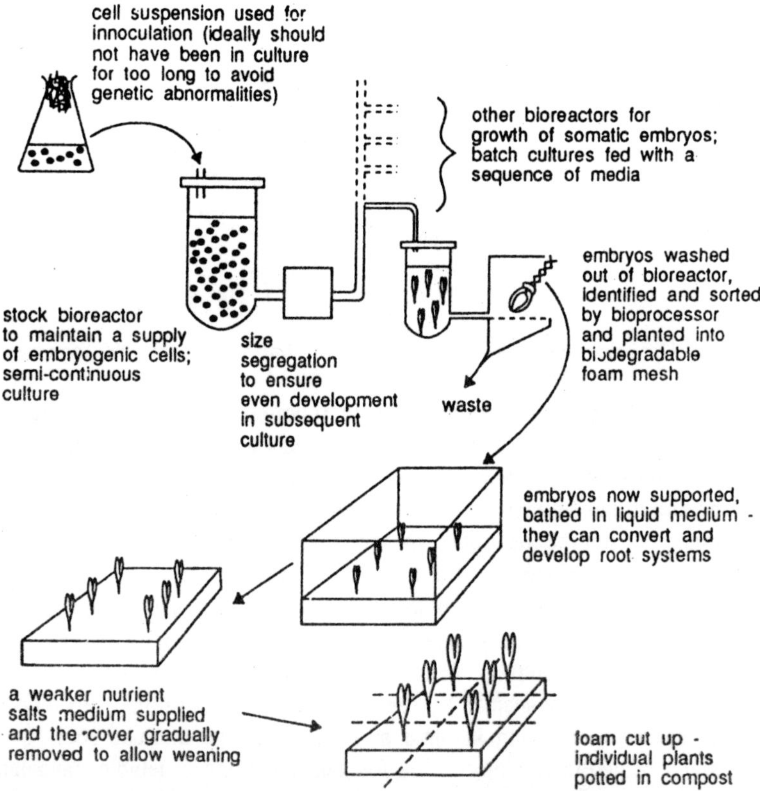

Fig. 12.6. A possible plan for linking bioreactors together to produce plantlets by somatic embryogenesis.

another. Thus many schemes suggest the use of some form of bioreactor for culturing and handling somatic embryos for micropropagation purposes. Unfortunately we are not yet able to control the growth of many species to make this a practical possibility. Work with model systems such as carrot and alfalfa is, however, well established, but no schemes for somatic seed production have yet become commercialized.

In the automation of micropropagation, the greatest difficulty arises when tissues must be divided up to permit their continued growth. As we noted above, somatic embryos have the ad vantage of being discrete units. When they arise from callus, merely shaking the container may be sufficient to dislodge them. In contrast, adventitious shoots must be cut apart with a sharp instrument. Simple as this task is to perform by hand, it can be difficult to mechanize (although great advances

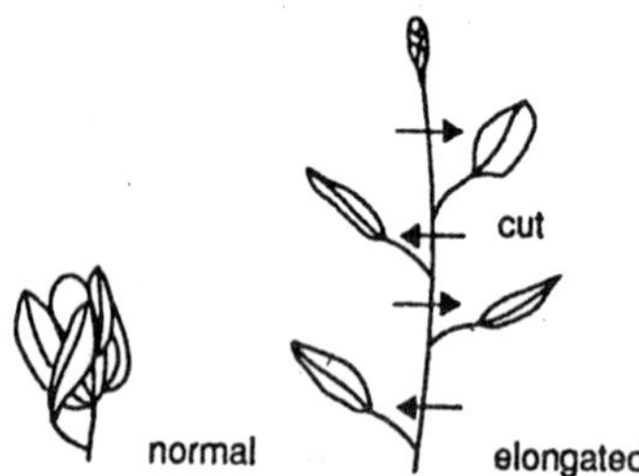

Fig. 12.7. How an elongated shoot can present a clearer picture to a robot (or a human) attempting to cut it into nodal sections.

have been made in recent years in robotics and machine vision). A mass of leaves and tangled shoots presents a difficult task for a robot to identify and to cut up into specific explants. The task can sometimes be simplified by causing culture conditions to elongate the tissues, but this may not be desirable in terms of quality of the plant product eventually produced by this form of microprogation.

Some species, notably ferns and some begonia cultivars respond well if coarsely homogenized at each subculture but most plants require more delicate handling. It may be possible to harvest shoots of pine and other woody species by repeatedly shaving them off from a hedge permanently growing *in vitro* and then inducing them to root. Quite sophisticated technology exists for gently holding vegetable transplants grown conventionally from seeds and attempts are being made to adopt this for handling tissues during micropropagation. It may be possible to use lasers to cut plant material accurately and without risk of contamination. Different morphological fragments of plants will respond to cutting up in varied ways. They may be identified and separated with the aid of a bioprocessor which incorporates many of the features described above.

Economic Considerations

At the start of this chapter we considered the reasons why micropropagation may be a desirable alternative to traditional horticultural propagating practice. In other parts of the world, the relative importance of these species is different—in the USA (where most micropropagation is carried out) the foliage plant *Syngonium* is most important.

There is considerable international trade in micropropagated plants. They may be sold by one producer to another at any stage in production

—in compost, bare-rooted or still in culture. Material *in vitro* at Stage II or III is particularly suitable for export since there are no risks of soil-borne pathogens being spread, however, some cultured tissues are not as disease free as they should be! Many companies use large shoot tip explants harbouring both viruses and bacteria which also get propagated in culture. Diseases of bananas and orchids which were formerly not a problem are now widespread as a result of micropropagation.

Table 12.3. Major tissue culture products

Product	*% of total*	*Product*	*% of total*
Gerbera	27	African violet	9
Ferns	18	*Ficus*	6
Lily	12	others	28

You must appreciate that micropropagation is more a commercial, rather than a scientific, process. Unless money can be made, the plants will be propagated by other means. Many small companies have been set up and failed because their production and marketing were inadequate.

Marketing

We said earlier that the massed cloning of new varieties of plants was one of the key functions of micropropagation. This is so, but it is relatively easy to saturate the market place with a particular ornamental plant. Many people buy such species as novelties—the plants will sell well at first, but when they become common, their value (and hence the profit for producing them) will fall. It is thus quite easy for a company suddenly to be left with a large (and perishable) stock of plants which must then be sold at a loss or even be discarded as unsaleable. A safer commercial position may be held if the company works on lower priced plants for which there is a continuous and predictable sale. It is for this reason that there has been interest in lowering production costs sufficiently to make the propagation of tropical plantation crops (such as banana and coffee) feasible.

Production

Much of this book describes how the culture environment may be modified to permit the optimal growth and development of the plant tissues. Exhaustive experiments to determine the best culture conditions are time consuming and therefore expensive. Even if optimum conditions are known, it may be impossible to provide them in a commercial situation. A huge range of lighting and temperature regimes is not practical; probably only two or three per factory can be used in practice. Similarly, although it is possible to tailor medium composition to different varieties of each species for the highest rates of plant production, in practice it is more satisfactory to use a basic range of media for as many species as possible. This cuts costs because larger batches of medium can be made with consequently less waste of surplus stock solutions.

Versatility of production is advantageous and is one of the benefits of a non-automated system. Human operators can be told quickly how to multiply a new species, but a robot may be more difficult to reprogramme. Current machine vision and robot manipulators are able to handle fewer plants per hour than their human counterparts. Although they do not need to be paid wages, their capital costs are high and research to develop them is too expensive for most commercial micropropagators without governmental sponsorship. Many ideas utilizing bioreactors and somatic embryos require sequences of media which must be modified for each different species or cultivar (a fact rarely stressed by their proposers!). While this is not an unsurmountable difficulty it does make the rapid changing of production from one genotype to another more difficult.

Determining Costs

Determining costs is essential when deciding whether to embark on propagating a new species or to set a price to ensure a profit. There are several methods which each have their merits:

1. The entire costs of the factory may be divided by the number of plants sold in a given period of time to estimate the cost of micropropagation;
2. Calculating the cost of operating a transfer cabinet per hour for each crop;
3. Calculating the cost per square metre of growth room shelf space. The productivity of this space will vary between crops;
4. Costs may be broken down to reveal which parts of the operation are most expensive;
5. The investment necessary to set up or expand a factory is considerable usually entailing the borrowing of large sums of money. Calculations of the increased productivity resulting from the investment will show how soon this cost can be repaid.

Micropropagation is a relatively new industry which is undergoing rapid change. The introduction of new methods permits a greater range of products to be offered at prices which can compete with established techniques and in many cases achieving a quality which was hitherto unattainable. Micropropagation is of critical importance since it provides an outlet for all the other areas of plant biotechnology. As a commercial undertaking, however, economic considerations must always be made before embarking on a new venture.

13

CELL SUSPENSION CULTURE

Cell suspension cultures are rapidly dividing suspensions of cells grown in liquid medium. In general, suspension cultures grow more rapidly than callus cultured on agar and are more amenable to experimental manipulation. The ideal suspension culture consisting entirely of single cells, which will allow the application of standard microbial culture techniques, is rarely ever achieved. Most suspension cultures are comprised of cell aggregates as well as dispersed single cells. However, by selecting and subculturing for several generations, a fine cell suspension culture consisting of a dispersion of single cells and small cell aggregates can be established and maintained.

INITIATION OF CELL SUSPENSION CULTURES

The time required to initiate and establish a cell suspension culture depends on the species of plant and the growth medium. However, in general, dicot species are easier to establish in cell suspension culture than monocot species.

Cell suspension cultures are usually initiated by agitating a fragment of *in vitro* grown callus in a volume of liquid medium on an orbital shaker. The essence of the procedure is to disperse callus into single cells and small masses of cells. Three main procedures are used to achieve dispersion:

- Initiation from friable callus;
- Initiation from non-friable callus;
- Initiation from callus treated with cell wall degrading enzymes.

In all cases, the callus selected for initiation of a suspension culture should be healthy and vigorously growing. The appearance of the culture is an important diagnostic feature; white or cream-coloured callus usually indicates a healthy culture whereas dark brown colouring suggests moribund or dead cells. To assess the condition of the culture at the cellular level, a little material may be taken aseptically and examined by light microscopy. The number of healthy cells can be quickly determined by staining with Evans Blue dye or the fraction of dividing cells can be estimated from the mitotic index by fluorescent microscopy with Hoechst stain. The most reliable measure of the viability and vigour of the callus culture is the growth rate.

Induction from Friable Callus

The most commonly used starting material for the initiation of suspension cultures is friable callus because it is easily fragmented during agitation in liquid culture. As friability is an important factor for the successful initiation of a suspension culture, a number of procedures can be used to obtain suitably friable callus. For example, the callus can be passaged on a 7-day cycle for 2-3 weeks immediately prior to its transfer to liquid culture and/or the ratio of auxin to cytokinin in the callus medium can be altered by raising the auxin concentration.

At initiation, it is also important to set up an appropriate ratio of callus tissue to liquid medium. Approximately 2-3 g of friable callus per 100 ml should establish a healthy suspension. Low levels of callus tissue will not replicate successfully. A denser seeding of callus can be monitored and passaged on at an early stage and is preferable to very low quantities of material in a large volume of liquid. As the cells break off from the callus and begin to form a suspension it will be necessary to subculture the cells to fresh medium; this should be kept to a ratio of 1:1. The procedure of repeated subculture should be continued until the culture reaches the desired density and is actively growing.

Actively dividing fine cell suspensions can be selected for at the early stages of culture initiation by filtering to remove the larger aggregates. By repeated filtering and subculturing, the culture can be reduced to a suspension of small aggregates and free floating single cells.

Initiation from Non-friable Callus

Where the callus available for the initiation of the cell suspension culture is non-friable, fragments of callus can be transferred to liquid medium and grown with agitation on a shaker. The callus is subcultured regularly until it reaches a suitable degree of friability. The callus is then used to initiate and establish cell suspension cultures as described above for friable callus.

Initiation from Callus Treated with Enzymes

In the early stages of initiation, callus is transferred and subcultured in a liquid medium containing low concentrations of cell wall degrading enzymes until a cell suspension is formed. Pectinase, which breaks down the middle lamella of the plant cell wall and separates plant cells, is frequently used and cellulase is sometimes added.

Maintenance of Cell Suspension Cultures

Cell suspension cultures can be grown asynchronously either in batch culture or in continuous culture. A variety of culture vessels and shakers are available commercially that can be used for the batch culture of plant cell suspensions. The most suitable vessel for batch is one that allows a large surface area to maximize gas exchange. For example, 20 ml of culture medium in a 100 ml conical flask or 50-100 ml in a 250 ml flask should allow sufficient gas exchange when the culture is being shaken at 100-120 rpm and at 25°C. Aluminium foil caps, sterilized and dried, should be used to seal the flask. Conical flasks with internal baffles are sometimes used to fragment any aggregates that might form. Established cultures are subcultured every 1-3 weeks depending on the growth of the culture. Transfers are made under aseptic conditions by using a wide-bore pipette to prevent blocking of the outlet. Suspension cultures should be subcultured in early stationary phase and at regular intervals, the embryogenic capacity of the culture should be assessed.

Several different types of fermentor or bioreactor can be used for the continuous culture of plant cell suspensions. The most common type of system for use in the laboratory is a stirred-jar fermentor as used for microbial cultivation, with a culture volume of around 7.5 l. Cell cultures grown commercially for the production of technical compounds and polymers are cultivated in sophisticated bioreactors with capacities of 75,000 l.

Growth Characteristics of Cell Suspension Cultures

Types of Cell

Plant cell suspension cultures normally consist of cells with diverse morphology and state of aggregation. Cells in aggregates exist in different microenvironments from single cells and respond differently to changes in the culture environment. In many suspension cultures, at least two morphological types of cells can be distinguished. Cell aggregates are normally made up of small isodiametric cells. Single cells may be large and elongated depending on the type of auxin used. The proportion of the cell types changes during the passage of the culture and depends on the nature of the auxin present in the culture medium.

Monitoring the Growth of the Culture

Growth of cell suspension cultures can be followed by measuring one or more of the following parameters at intervals during the growth cycle:

- Cell number;
- Packed cell volume (PCV);
- Fresh weight and dry weight;
- Cell viability;
- Medium conductivity.

It is good practise to choose two methods and to use these to measure the culture until it appears to have entered stationary phase.

Cell number

The number of cells per unit volume can be determined using a haemocytometer. The suspension is diluted so that at least three fields can be scored and the total cells recorded should be greater than 1000. It should be noted that cell aggregates and the large size of single cells in some suspensions can cause incomplete loading of the counting chamber of the haemocytometer and consequently lead to unreliable cell counts. These difficulties can be overcome by dispersing the cell aggregates with 10% HCl and 10% chromic acid as described by Reinert and Yeoman (1982) and by using an improved Neubauer type haemocytometer (depth of counting chamber, 0.1 mm) to accommodate the larger cells. Great care should be taken when using corrosive acids and the acid treatments should be carried out in a fume hood.

Packed cell volume (PCV)

Aliquots (10-15 ml) of evenly mixed cell suspensions are transferred to a graduated tube and spun at low speed in a centrifuge fitted with a swing-out rotor. The PCV is calculated as the percentage total volume occupied by the cell pellet. The pellet can be macerated and the cell count determined as described above. PCV is accurate only for fine suspension cultures.

Fresh weight and dry weight

Fresh weight is determined by weighing freshly harvested cells and dry weight from cells dried at 60°C for 48 *h*.

Cell viability

Rough estimates of the percentage of viable cells in a culture can be made frcm cell counts by using viability stains like Evans blue and fluorescein diacertate.

Conductivity and pH

The conductivity of the growth medium is sometimes used as a convenient method of monitoring the growth of a suspension culture. Conductivity measurements can be made rapidly with a conductivity meter, but each new suspension culture must be assessed for the accuracy of this technique by comparison with other growth data. In general, the conductivity of the media falls over the passage and probably reflects the uptake and utilization of electrolytes from the medium.

Changes in pH also reflect the various phases of the culture cycle. Cultures become very acidic early on in logarithmic phase (pH 3 5-4.5) and return to approximately pH 5 in stationary phase. The pronounced drop of pH early in the growth cycle may be due to proton production from metabolic activity and the rapid removal of media constituents like phosphate that provide the buffering capacity of the medium.

CONTAMINATION

It is essential to check cultures before and soon after subculturing for any signs of microbial contamination. An opaque, dense white or pink colouration that appears very rapidly (i.e. within 24 h) is usually indicative of a yeast or bacterial contamination. The identity of the contaminant can be confirmed by examining a sample of the culture under the microscope. Fungal contamination usually appears as balls of mycelium that develops 48 h after transfer. All cultures suspected of contamination should be disposed of immediately by autoclaving.

USES OF CELL SUSPENSION CULTURES

Plant cell suspension cultures have found wide application both for research and for commercial exploitation. For research, cell suspension cultures are easy to maintain and allow great flexibility of experimental approach. Cell suspensions are ideal to study various factors and compounds that affect growth and differentiation. One of the key advantages is the feature that most of the cells are in direct contact with the medium and therefore the effects of concentration gradients are avoided. Cell suspension cultures provide a useful system to study cell division in the absence of developmental processes and their growth can be made synchronous for studies on the cell cycle. Another important feature of suspension cultures is that they can be used for the rapid preparation of protoplasts in high yield. Protoplasts themselves are a unique system for basic and applied research.

The formation of somatic embryos in suspension cultures is ideal for the large-scale production of commercial plants. The greatest potential commercial application of suspension cultures is their use as a production system for plant-derived chemicals and recombinant pharmaceutical proteins.

PROTOCOL

Generic Protocol for Initiating a Suspension Culture

This is a protocol suitable for different types of callus

Equipment

Incubator set at correct temperature for callus

Orbital incubator, shaking at 130 rpm, set at correct temperature for species, fitted with fluorescent lights and timer

Laminar flow cabinet

Bunsen burner

Autoclave

Scalpel and forceps, autoclaved in aluminium foil

Glass Petri dish (to hold tools while cooling)

Sterile glass or plastic Petri dishes.

Materials and reagents

Petri dishes of solid medium containing calli to be transferred to suspension cultures

Flasks of medium appropriate to the species of calli. Volumes of 20 ml in a 100 ml flask or 50-100 ml in a 200 ml flasks are recommended

70% w/v alcohol.

Procedure

1. Thoroughly sterilize the laminar flow cabinet with 70% alcohol. Leave running for 20 min before use.
2. Spray foil-wrapped tools with 70% alcohol and leave to dry in the cabinet

3. When bringing plates of callus into the cabinet, check carefully that the marker pen used to label the plates does not wash off in the presence of alcohol. If it does, wipe unmarked areas of plates with paper soaked with alcohol.
4. Spray flasks of medium with 70% alcohol and allow to dry.
5. Unwrap sterile toots and Petri dishes of callus.
6. Using forceps and scalpel move callus to sterile Petri dish and break apart into small pieces. Ideally, the tissue should be approximately 2-3 mm in diameter. Close Petri dish.
7. Resterilize the tools in a bead sterilizer. Allow to cool before use.
8. Remove foil cap from fresh flask of medium.
9. Using forceps, pick up callus and transfer into flask. Avoid touching sides of neck with callus.
10. Seal neck of flask with a fresh foil cap.
11. Move flask to orbital shaking incubator. Monitor growth of material by eye. Discard contaminated flasks. Within a week or two it should be possible to observe an increase in the number of free cells in the medium which have broken away from the callus.
12. After 7-10 days, the culture should be passaged, whether or not a large increase in mass has occurred. If the volume of tissue has not increased greatly, use a smaller volume of medium for the second passage.
13. Allow the material to settle at the bottom of the flask. Pipette off the liquid medium and replace with fresh.
14. Repeat this method of subculture until a reasonable mass of material his accumulated. Then pipette a volume of cells (for example 10 ml into a 100 ml) from the old flask into a fresh flask.
15. Should the material not be increasing in mass at an appreciable rate, repeat the procedure from the beginning using a larger volume of dispersed callus to the same volume of liquid medium.

Monitoring the Growth of Suspension Cultures

Part A—Packed cell volume (PCV)

Equipment

Sterile 10 ml pipette
Automatic pipette (pipette filler)
Sterile, graduated plastic centrifuge tube
Bench centrifuge (swing-out rotor preferable)
Graph paper.

Materials and reagents

Flasks with equal volumes of suspension culture cells, subcultured from a single source flask. There should be one flask for each day of the planned time course.

Procedure

1. Pipette a fixed volume of suspension culture into a 15 ml graduated plastic centrifuge tube and either allow cells to settle out by gravity or centrifuge at 1000 g for 5 min. Ensure the flask is thoroughly shaken immediately before sampling as suspension culture cells settle very rapidly.
2. Record the volume of the pellet.
3. Repeat this every day at the same time and plot against time. The numbers should show a rapid increase during the logarithmic phase of growth and then reach a plateau. This should be repeated for the new culture when its growth cycle is stable. With this information, establish a day when the culture is in early stationary phase and choose this as the routine day for subculturing.

Part B—Measurement of fresh weight and dry weight

Equipment

Glass filter funnel

Whatman no. 1filter paper, 9 cm in diameter

Buchner flask (with side arm for vacuum attachment)

Fine balance

Oven set to 60°C.

Materials and reagents

Flasks of cells. There should be a separate flask of cells for each day of the planned time course.

Procedure

1. To determine wet weight, filter the cell suspension through a pre-weighed Whatman no. 1 filter paper, allow it to drain under vacuum and reweigh the filter.
2. To determine a dry weight, use a pre-weighed filter paper, filter a known volume of cells through it and dry in a 60°C oven for 48 h. Because dried cells weigh very little this method is likely to be inaccurate unless large volumes are filtered.
3. Repeat with replicates to obtain an accurate growth curve. This should be repeated at the same time every day until no further growth is detected and the culture has reached stationary phase.

4. Plot the weights, wet or dry against time to produce the growth curves.

Measurement of Cell Viability

Part A—Exclusion of Evans blue dye

Equipment

Pipette

Light microscope

Small plink aim. 1.5 ml in capacity

Slides and coverslips.

Materials and reagents

Flask of suspension-cultured cells to be assayed

Stock solution of Evans blue stain, 0.1% w/v in water.

Procedure

1. Mix 1 drop of stock Evans blue dye solution with 0.5 ml of cell suspension in a small plastic tube.
2. Allow to incubate for 2 min at room temperature.
3. Transfer 1 drop to a slide and apply a coverslip, taking care to avoid bubbles.
4. View under the light microscope. Ensure that there is a minimum of 100 cells in the field of view. Living cells exclude Evan's blue dye, while dead cells are stained blue.
5. Count the numbers of stained and unstained cells. Calculate the percentage of living cells as a measure of viability.

Part B—Fluorescent staining with fluorescein diacetate (FDA)

Equipment

Fluorescence microscope

Slides and coverslips

Automatic pipettes, 0-10-μl and 200-1000-μl capacity

Eppendorf tube, 1.5 ml capacity.

Materials and reagents

FDA stock solution prepared as 0.5 mg ml^{-1} dissolved in acetone.

Cell or protoplast suspension.

Procedure

1. Mix 5 μl of stock FDA solution with 0.5 ml of protoplast suspension in an Eppendorf tube.
2. Incubate for 2 min at room temperature.

3. Place a drop on a slide, cover with a coverslip and view under a fluorescence microscope. Ensure that there are at least 100 cells visible and that they are separate from each other.
4. Count the fluorescent and non-fluorescent cells in the field of view. The fluorescent cells are viable. The percentage of viable cells should exceed 80% in a healthy suspension culture.

Part C - Mitotic index with Hoechst stain

Equipment

Fluorescence microscope

Slides and coverslips

Automatic pipettes 0-10 μl and 200-1000 μl capacity

Materials and reagents

Hoechst stain 1 mg ml^{-1} in water (wear gloves when handling)

Protoplast or cell suspension.

Procedure

1. Add 1 drop of Hoechst stain to a drop of cells on a microscope slide.
2. Incubate for 5-15 mins.
3. Apply coverslip and gently absorb excess stain with tissue.
4. Observe by fluorescence microscope with U.V filter set. Dividing nuclei will fluoresce strongly blue. Mitotic index can be estimated from the proportion of cells undergoing division.

Genetic Transformation of Suspension Cultured Cells

Equipment

Small Petri dish

Automatic pipette and sterile tips

Incubator, without lights, set at 25°C

Bench centrifuge and tubes

Sterile forceps and scalpel.

Materials and reagents

Flask of suspension cultured cells, 3 days old

Agrobacterium tumefaciens overnight culture (OD at 600 nm reading 2.0)

Fresh, sterile cell culture medium

Petri dishes of cell culture medium plus 0.8% agar and supplemented with a selecting agent such as an antibiotic (e.g.

kanamycin or hygeromycin) and an antibiotic to kill the Agrobacteria (e.g. carbenicillin and timentin).

Procedure

1. Take 1 ml of a 3-day-old cell suspension of cells and transfer to a small, sterile Petri dish.
2. Add 50 μl of an overnight culture of *Agrobacterium tumefaciens*.
3. Mix the culture and incubate in the dark for 2 days at 25°C.
4. Aseptically transfer by pipette to a centrifuge tube and spin at 50 g for 5 min.
5. Wash pellet three times with sterile culture medium to remove excess, unattached bacteria.
6. Take plates of solid culture medium supplemented with appropriate selecting antibiotic and antibiotic mix to remove *Agrobacteria*.
7. Aseptically pipette 1 ml of the culture on to each plate and spread cells across the entire surface by shaking.
8. Incubate for 1 month at 25°C in the dark to allow microcalli to develop on the plates.
9. After this time, screen the plates for transformed microcalli (for example, if using GFP expression, screen with an ultraviolet lamp).
10. Transfer the selected microcalli, using sterile forceps, on to fresh plates.

14

Bacterial Culture

The production of transgenic plants requires methods for identifying, isolating, replicating and inserting DNA into the plant genome. All these steps may involve the use of bacteria and bacterial culture (especially of *E. coli* and *Agrobacterium tumefaciens*) is routine in the plant tissue culture laboratory undertaking plant transformation. Thus while the culture techniques required are different from those for plant and cell culture, a basic description will be given here.

A vector (which may be either bacterial or viral) is needed to introduce a gene into a foreign genome. The vector system most commonly used for plant transformations is based on the tumour-inducing plasmid of *A. tumefaciens*. Prior to insertion into a transformation vector, the isolated gene must be integrated into a cloning vector—a bacterial plasmid which can be multiplied in a bacterium like *E. coli* that is easy to handle in the laboratory. *E. coli* and *A. tumefaciens* are available as modified strains that minimize risk to the environment. *E. coli* cells treated with calcium chloride will take up plasmids from the culture medium and replicate them. They can be grown to logarithmic phase, incubated in calcium chloride and snap-frozen as competent cells—ready to take up a plasmid when required. Fragments of DNA can also be packaged into bacteriophage λ, which can be used to transform the bacteria.

Agrobacterium rhizogenes is also used as a transformation system for plant cell cultures, where it produces 'hairy' roots—roots with abundant root hairs. These have been used in tissue culture as they are effectively immortal and produce abundant and rapidly growing root cultures. Techniques for the culture of *A. rhizogenes* are therefore also included with those for *E. coli* and *A. tumefaciens*.

Facilities for Bacterial Culture

To culture bacteria successfully, a clean bench in a room that is not a through room is recommended. A Bunsen burner can be used to sterilize tools and to flame the tops of vessels before transfer takes place. It also serves to provide a circular area on the bench that can be regarded as sterile to work in. Glass spreaders should be dipped in 70% alcohol and held in the Bunsen flame until all the alcohol has burnt off to sterilize them. When holding these it is important to hold the head of the spreader at a lower position than your hand so that burning alcohol cannot flow on to you. These must be cooled before use. To sterilize a wire loop, it should be heated until redhot before cooling. In contrast to the spreader, the head of the loop should be held up in relation to the hand so that the handle does not heat up and burn the operator. An alternative to using loops to inoculate a liquid culture is to use an automatic pipettor and sterile tips. Small, open containers lined with an autoclave bag may be used for disposal of these tips. Both the loop and the glass spreader should he sterilized by Bunsen immediately after use.

To culture bacteria, both a stationary incubator for plates and an orbital, shaking incubator for liquid cultures, set at 28°C for *Agrobacteria* or 37°C for *E. coli*, are necessary. For bulking up bacterial cultures, orbital shakers that are capable of holding 2-1 conical flasks should be available. Most modern platform shakers are flexible enough to cope with a range of sizes and shapes of vessels by supplying racks and clamps. Bacterial cultures do not require illuminated growth facilities. To deal with solid, contaminated waste, a metal bucket suitable for the autoclave should be under the bench. Plastic boxes, also suitable for the autoclave are available. Both containers should be lined with an autoclave bag before use. For liquid-contaminated waste, a jug of hypochlorite solution or disinfectant like Virkon™ should be on the bench or on a low platform very close by. These should be renewed daily. Powdered Virkon™ and a source of paper towel should be close at hand in case of a liquid spill. It is essential to have access to an autoclave in order to sterilize your media and equipment, and also to dispose of contaminated waste, but it should not be inside the culture facility.

Other general equipment for bacterial culture includes a pH meter, source of ultrapure water, magnetic stirrer, balance, vortex mixer and cuvettes for spectrophotometry (glass or disposable plastic for use at visible wavelengths and quartz for UV spectra). Gloves and water and alcohol-proof marker pens are also essential. A microwave oven can be used to melt agar stocks and a water bath set at 45°C can be used to maintain molten agar until it is poured. However, never place sealed containers in a microwave oven; even containers with a lightly screwed lid can seal and generate explosive pressure.

PROTOCOL

To Culture *E. coli* in Luria Broth (LB)

Equipment

Bacterial wire loop
Bunsen burner
Shaking incubator set at 37°C
Spectrophotometer set at 600 nm
Timer
Sterile glass universal bottles and tops
Sterile 250 ml Erlenmeyer flask sealed with cotton wool.

Materials and reagents

70% alcohol to clean the bench
Sterile Luria broth: tryptone, 10 g l^{-1}; yeast extract, 5 g l^{-1}; NaCl, 10 g l^{-1}; D-glucose, 10 g l^{-1}, pH 7.0
Plates of LB agar (1%) streaked with the *E. coli* strain of choice 24 h prior to use.

Procedure

1. Sterilize wire loop by heating in Bunsen flame until it is red hot. Allow to cool.
2. Inoculate a single colony by loop from a fresh plate of the *E. coli* strain into 10 ml of LB and grow overnight at 37°C in a shaking incubator at 150 rpm.
3. In the morning, inoculate 500 μl of overnight culture into 50 ml of fresh LB and grow for 90 min to an optical density of 0.1 at 600 nm. These should be in the early phase of logarithmic growth and will be ready for transformation or for further use.

Calcium Chloride-mediated Transformation of *E. coli*

Equipment

Bench centrifuge and centrifuge tubes
Glass spreader
Glass universal bottles and 200 ml flasks
Ice in ice bucket

Timer
Micropipettor and tips
Pipettes
Shaking incubator set at 37°C
Spectrophotometer
Water bath at 42°C.

Materials and reagents

Antibiotic for selection of *E. coli* transformed with plasmid
100 mM $CaC1_2$
50 mM $CaC1_2$
E. coli strain (suitable for high-frequency transformation, e.g. DH5a)
Ethanol LB
LB supplemented with 1% agar
Plasmid DNA
Tris-EDTA buffer (TE): 10 mM Tris-HCl, pH 7.5, 1 mM EDTA.

Procedure

1. Inoculate a single colony from a fresh plate of the *E. coli* strain into 10 ml of LB and grow overnight at 37°C in a shaking incubator.
2. Inoculate 500 μl of overnight culture into 50 ml of fresh LB and grow for 90 min, to an absorption at 600 nm of approximately 0.1.
3. Centrifuge the cells in a cooled centrifuge for 10 min at 4000 g.
4. Pour off the supernatant and gently resuspend the pellet in 15 ml of ice-cold 100 mM $CaCl_2$. Incubate the cells on ice for 30 min. *E. coli* cells rapidly lose their competence to take up DNA if they are kept warmer than 4°C, or if they are not treated gently.
5. Centrifuge the cells at full speed in a cooled bench centrifuge, pour off the supernatant and gently resuspend the pellet in 2 ml of ice-cold 50 mM $CaCl_2$. (At this stage of the procedure, the cells are described as competent. They may either be snap-frozen in liquid nitrogen and stored for up to 3 months or they may be left on ice for a minimum of 2 h then transformed as in the next steps.)
6. Add up to 100 μl of plasmid DNA to 200 μl of competent cells and incubate on ice for 30 min.
7. Heat-shock the cells by placing the tubes in a 42°C water bath for 2 min. Return the tubes to the ice bucket and allow to recover for 5 min.

8. Transfer the cells to a glass universal bottle. Add 1 ml of LB and grow for 1 h at 37°C in a shaking incubator.
9. Prepare LB agar plates with the appropriate antibiotic to select for the transformed *E. coli* cells.
10. Plate 50 μl of transformed cells on to the selective plate and spread smoothly, using a flamed glass spreader.
11. Incubate the plates at 37°C overnight or until the colonies are visible.
12. Pick single, well-isolated colonies and check for the presence of the plasmid using standard molecular biology techniques.

Calcium Chloride-mediated Transformation of *Agrobacterium*

Equipment

Bench centrifuge and centrifuge tubes
–70°C freezer
Ice in an ice bucket
Automatic pipettor and sterile tips
Shaker set at 28°C
Spectrophotometer
Water bath set at 37°C
Sterile freezing tubes.

Materials and reagents

Agrobacterium strain of choice
Antibiotics for selection of binary construct
20% v/v glycerol
Liquid nitrogen
TE medium: 10 mM Tris-HCl, pH 8.0, 1 mM EDTA
TY medium: 0.5% tryptone, 0.3% yeast extract, 0.13% $CaCl_2$, pH 7.0
TY agar:TY medium supplemented with 1% agar
Vector DNA from binary construct.

Procedure

1. Inoculate a 10 ml culture of TY medium from a single colony of *Agrobacterium* and grow in a shaking incubator at 28°C overnight.
2. Inoculate 200 μl of the culture into 200 ml of TY medium and shake at 28°C until the culture reaches an absorbance of 0.4 at 600 nm (8-12 h).
3. Centrifuge the culture in a cooled centrifuge at 4000 g for 10 min at 4°C. Pour off supernatant and wash pellet with 20 ml of TE.

4. Repeat centrifugation step and resuspend the pellet in 20 ml of ice-cold TY medium.
5. Dispense the cells into 500 μl aliquots in 1.5 ml freezing tubes and immediately freeze in liquid nitrogen. Care must be taken in using liquid nitrogen. At this state, the competent cells may be either frozen in a –70°C freezer and stored for up to 3 months, or thawed on ice and transformed as in the subsequent procedure.
6. After thawing, add approximately 1 μg of binary construct DNA in a volume not exceeding 50 μl and mix gently.
7. Incubate the cells for 5 min on ice.
8. Transfer the tubes to liquid nitrogen for 5 min and then plunge them into a 37°C water bath for 5 min.
9. Add 1 ml of TY medium and shake the cultures at 28°C for 4 h.
10. Plate the transformed cells on TY plates containing antibiotics for the selection of the binary construct. Grow at 28°C for 24-48 h.
11. Pick single colonies and restreak on to new plates.
12. Store the resulting strain by resuspending bacteria from the plate in sterile 20% glycerol and freezing at –70°C.

Culture of *Agrobacterium tumefaciens* in Culture Medium, YEB

Equipment

Incubator set at 28°C

Shaking incubator set at 28°C

Bacterial wire loop

Spectrophotometer set to 600 nm

Bench centrifuge and centrifuge tubes.

Materials and reagents (all sterile)

YEB: 10 ml aliquots sterilized in glass universal bottles; yeast extract, 1 g l^{-1}; beef extract, 5 g l^{-1}; peptone, 5 g l^{-1}; sucrose, 5 g l^{-1}; $MgSO_4.7H_2O$, 0.5 g l^{-1}; pH 7.0

YEB: 50 ml sterilized in 250 ml Erlenmeyer flask, sealed with cotton wool

YEB agar plates

Antibiotics.

Procedure

1. Inoculate a single colony of the *A. tumefaciens* strain from a fresh plate into a 10 ml culture of YMA liquid medium supplemented with antibiotics appropriate to the strain and construct resistance.

2. Grow overnight at 28°C in a rotary shaker at 130 rpm.
3. Next morning, inoculate 500 μl of the overnight culture into 50 ml of fresh YEB without antibiotics.
4. Grow this overnight at 28°C, shaking at 130 rpm until the culture is in the late logarithmic phase. The optical density at 600 nm should be around 0.8.
5. This culture can be centrifuged at 5000 g for 10 min, washed in appropriate plant culture medium, resuspended to an O.D. of 2.0 and used for transformation of leaf discs.
6. *A. tumefaciens* is sensitive to the antibiotic carbenicillin and the combination antibiotic timentin (clavulanic acid/ticarcillin).

Culture of *Agrobacterium rhizogenes* in Culture Medium, YMA

Equipment

Incubator set at 28°C

Sterile syringe needle.

Materials and reagents

YMA broth plus 1% agar plates: $MgSO_4.7H_2O$, 2g l^{-1}; K_2HPO_4, 0.5 g l^{-1}; NaCl, 0.1 g l^{-1}; mannitol, 10 g l^{-1}; yeast extract, 0.4 g l^{-1}; pH 7.0

Antibiotics

Plant tissue.

Procedure

1. Inoculate the chosen *Agrobacterium* strain on to a fresh plate of YMA plus antibiotics appropriate to the strain and incubate overnight at 28°C.
2. When using to transform plant tissue, dip the tip of a sterile syringe needle into the culture of *A. rhizogenes* and use to stab the sterile plant tissue at 2-3-mm intervals.
3. The infected plant material should be incubated under conditions suitable for the plant tissue. Hairy roots should emerge within 14 days.
4. *A. rhizogenes* is sensitive to the antibiotic Cefotaxime.

15

Somaclonal Variation

The culture regimes used for the regeneration of whole plants from tissue explants, undifferentiated cells and protoplasts are frequently a source of variability. This means that a significant percentage of the regenerated plants may not be identical in genotype and phenotype to the plant from which the original explant was obtained. Larkin and Scowcroft (1981) named this phenomenon somaclonal variation.

The origin and expression of the observed variation differs from species to species as well as the specific sources of the tissue explants.

Origins and Mechanisms of Somaclonal Variability

Somaclonal variation can be of two sorts:

1. Genetic (i.e. heritable) variability—caused by mutations or other changes in DNA;
2. Epigenetic (i.e. non-heritable) variability—caused by temporary phenotypic changes.

Genetic Variability

Various molecular mechanisms are responsible for genetic variability associated with somaclonal variation.

Changes in ploidy

One of the more frequently encountered types of somaclonal variation results from changes in chromosome number, that is, aneuploidy, polyploidy or mixoploidy. Changes in ploidy originate from abnormalities that occur during mitosis. For example, extra chromosomal duplication during interphase, spindle fusion or lack of spindle formation and cytoplasmic division.

As plant cells grow and age, the frequency of changes in ploidy increases. Therefore, changes in ploidy observed in cultures and regenerated plants might have their origins in the source of tissue explants used.

Another cause of variability due to changes in ploidy is the *in vitro* culture regime itself. The longer the cells remain in culture the greater is their chromosomal instability. In addition, the composition of the growth medium can trigger changes in ploidy. For example, both kinetin and 2,4-D are implicated in ploidy changes and cultures grown under nutrient limitations can develop abnormalities. Selecting a suitable explant and an appropriate culture medium can therefore enhance the chromosomal stability of the culture. However, high variations of ploidy in cultures do not always lead to high frequencies of somaclonal variation in regenerated plants. This is because, in mixed cultures, diploid cells appear to be better fitted than aneuploid or polyploid cells for regeneration, as they are more likely to form meristems.

Structural changes in nuclear DNA

Structural changes in nuclear DNA appear to be a major cause of somaclonal variation. The changes can modify large regions of a chromosome and so may affect one or several genes at a time. These modifications include the following gross structural rearrangements:

1. Deletions—loss of genes;
2. Inversions—alterations in gene order;

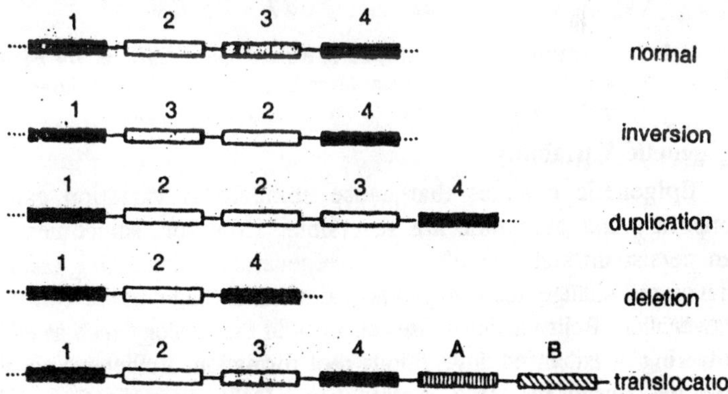

Fig. 15.1. Patterns of gross DNA changes that can cause somaclonal variation. Product 1 is generated by the arbitrary primers and product 2 by primers B and D. Products from the PCR reaction are separated by electrophoresis and the pattern further analyzed.

3. Duplications—duplication of genes;
4. Translocations—segments of chromosomes moving to new locations.

Activation of transposons can be a cause of somaclonal variation. Transposons or transposable elements are mobile segments of DNA that can insert into coding regions and cause gene disruption.

In addition to these larger modifications of nuclear DNA sequence, changes at the level of a single DNA nucleotide that occur in a coding region can lead to somaclonal variation. For example, point mutations that result from a change of base in a single nucleotide or the altered methylation of a base can lead to gene inactivation.

Chimaeral rearrangement of tissue layers

Many horticultural plants are periclinal chimaeras, that is, the genetic composition of each concentric cell layer (LI, LII, LIII) of a meristem (e.g. the shoot tip meristem) is different. These layers can be rearranged during rapid cellular proliferation. Therefore, regenerated plants may contain a different chimaeral composition or may no longer be chimaeric at all. Shoot tip transformation procedures are particularly likely to cause chimaeral transgenics.

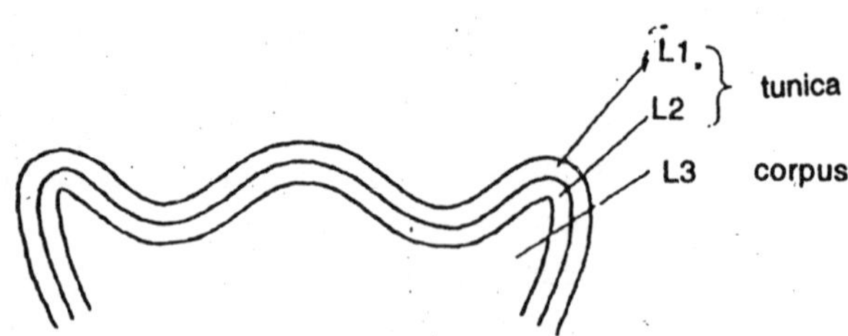

Fig. 15.2. The organization of layers L1, L3 in a shoot apical meristem. The meristem is viewed in transverse section; layer L1 forms the epidermis, layer L2 the subepidermal layer and layer L3 the internal tissues of shoot.

Epigenetic Variability

Epigenetic changes that cause somaclonal variation can be temporary and over time are reversible. However, sometimes they can persist through the life of the regenerated plant. One common phenotypic change seen in plants produced through tissue culture is rejuvenation. Rejuvenation causes changes in morphology such as earlier flowering or enhanced adventitious root formation. Epigenetic changes may be caused by DNA methylation, DNA amplification or by activation of transposable elements (transposons). DNA methylation is important for both transcription and translation because it alters the substrate affinities of enzymes that are active in these processes. Tissue culture media can change the level of DNA methylation and thus may be one of the important causes of somaclonal variation.

Importance of Somaclonal Variation

The *in vitro* procedures for transformation, regeneration and clonal propagation of plants may involve tissue culture procedures that can take anywhere from 4 weeks to several months to complete. The degree of somaclonal variation generated and accumulated during this period of culture can be significant and is an unwelcome interference that is difficult to control. In plant transformation procedures, apart from the mutagenic effects of the basic *in vitro* culturing process, methods of DNA delivery and selection may also induce additional somaclonal variation.

Extensive somaclonal variation can result in undesirable changes in the genetic background of regenerated plants that could limit their use as cultivars or as elite parents in breeding programmes. It is therefore important to assess the genetic variability induced during transformation and/or regeneration of plants through tissue culture.

Methods of Assessing Somaclonal Variation in Regenerated Plants

Although cytological and phenotypic analyses can be used to evaluate somaclonal variation, recently molecular techniques have been used with increasing frequency.

Restriction Fragment Length Polymorphism (RFLP)

RFLP was one of the first techniques to be applied to somaclonal variation and has been widely used for several species. RFLP is a hybridization-based technique that detects variation at the DNA sequence level but requires the use of probes that hybridize to known sequences.

A number of amplification techniques based on PCR technology have now been developed that avoid the need for prior sequence information.

Random Amplified Polymorphic DNA-PCR (RAPD-PCR)

RAPD-PCR or arbitrarily primed PCR (AP-PCR) is a technique that has proved useful in detecting somaclonal variation in a number of species. RAPD-PCR is based on the premise that, because of its complexity, eukaryotic nuclear DNA may contain paired random segments that are complementary to single decanucleotides and furthermore these segments have the correct orientation and are located

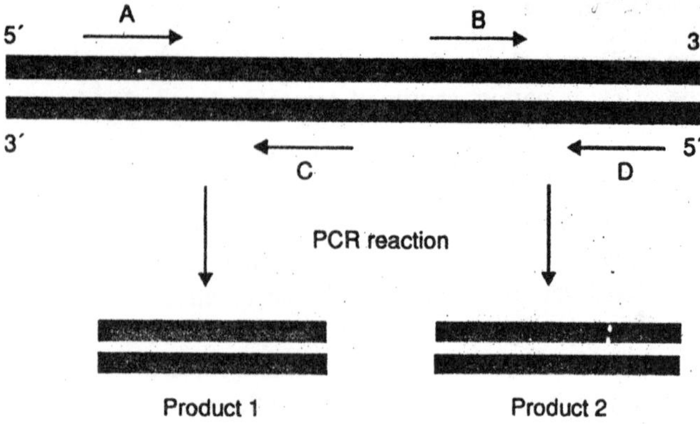

Fig. 15.3. Illustration of the principle of RAPD.

close enough to each other for PCR amplification. RAPD-PCR uses single primers of arbitrary nucleotide sequence to initiate DNA synthesis. The DNA fragments can be separated by gel electrophoresis and the DNA variation is detected by the pattern of DNA bands from individual plants.

Amplified Fragment Length Polymorphism (AFLP)

More recently, another,PCR-based technique known as AFLP has been used to study somaclonal variation. AFLP is a DNA-fingerprinting procedure based on the selective PCR amplification of fragments from restriction digestion of gnomic DNA.

AFLP analysis consists of the following steps.

1. Genomic DNA is digested with two restriction enzymes, one that cuts frequently, for example *Msel* (4-bp recognition sequence) and one that cuts less frequently, for example *Eco*RI (6-bp recognition sequence).
2. The resulting fragments are ligated to double-stranded adapter molecules that consist of a core sequence and a sequence specific for either the *Eco*RI site or the *Mse*l site.
3. Pre-selective amplification by PCR using primers designed to include a core sequence, an enzyme specific sequence and a single base extension at the 3' end. The primary products are fragments containing one *Msel* cut, one *Eco*RI cut and a matching internal nucleotide. This amplification step achieves a 16-fold reduction in complexity.
4. Selective PCR amplification using the products of the pre-selective amplification as template and identical primers as the pre-selective

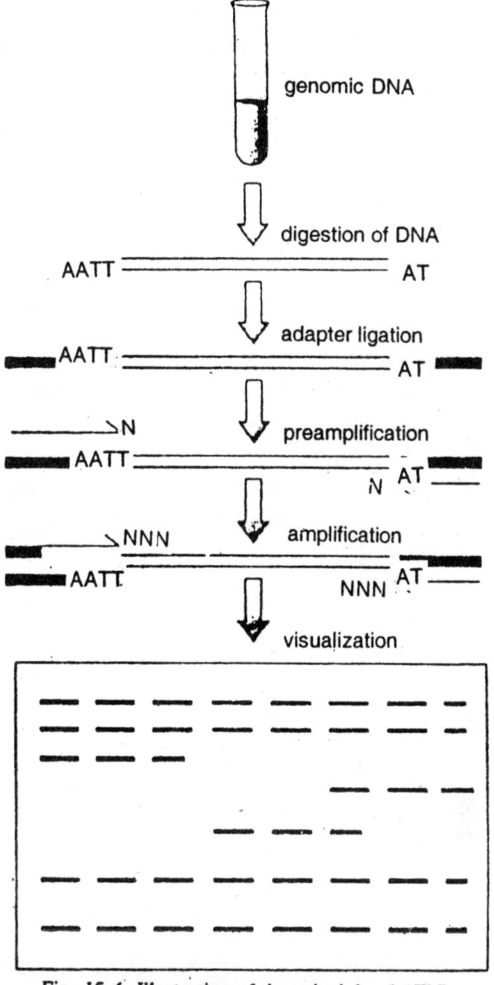

Fig. 15.4. Illustration of the principle of AFLP.

step, but containing two further additional nucleotides at the 3' end. The primers are either radio-labelled or fluorescently labelled. Only fragments having the matching nucleotides in all three positions (50-200) will be amplified. This step reduces the complexity 250-fold.

5. Gel electrophoretic separation reveals the pattern (fingerprint) of the labelled fragments that are analysed with the aid of an appropriate software package.

The molecular analyses described above are rapid and more precise the phenotypic. analyses. However, both the RAPD and AFLP techniques have proved to be inconclusive in some studies.

Somaclonal Variation as a Technique for Crop Improvement

The recognition that cell culture regimes were sources of somaclonal variation in regenerated plants led to the use of cell culture for mutagenesis and the direct selection of genetic variants with valuable traits. Such a selection strategy was considered a particularly effective way of obtaining variants that were tolerant to environmental factors such as cold, drought, salinity or toxic ions or resistance to pathogens. Although there are cases where somaclonal variation has produced plants with useful agricultural traits, as a source of variation somaclonal variation is difficult to exploit because the variation is largely unstable, mostly unpredictable and can be deleterious. This technique has not proved useful as a source of genetic variation for plant improvement as was once anticipated.

16

Tissue Culture in Agriculture

Modern humans began farming about 10,000-12,000 years ago. This marked a shift from gathering food and fibre from wild plants, to the deliberate cultivation of crops. The transition was probably gradual but was one of the most dramatic changes in the history of mankind. As a result of settled agriculture, humans began to select, safeguard and propagate variants of species that had the most desirable phenotypic characteristics. The activities of planting, harvesting and storing seed for replanting and propagation imposed selection pressures. Over several millennia, these selection pressures resulted in changes in the phenotypes of wild plants, which led to the phenotypes that are characteristic of their domesticated crop relatives. With time, the farmer-selected crops became adapted to the local conditions and were maintained by man as 'primitive cultivars' or land races. Starting with land races, farmers over centuries used simplistic breeding methods to try to improve crops and to create plants that would grow effectively in regions of the world far beyond their centres of origin and domestication. However, the process of crop improvement by these methods was slow and it was not until the beginning of the 20th century that rapid progress was made, when sound scientific method was introduced to plant breeding by the application of Mendelian genetics. For the next four or five decades up until the early 1970s, methods of breeding for specific traits were developed that made significant contributions to agriculture, by delivering improvements in crop yield and enhancing plant resistance to pest and diseases.

Plant breeding depends on the sexual recombination of selected plants to generate new genotypes that can be propagated through seed.

Over the centuries, in parallel with the evolution of plant breeding methods, techniques were developed for the asexual (vegetative) propagation of selected plants. In asexual propagation, the new plants are exact copies or clones of a single parent plant. Classical horticultural methods of clonal propagation include budding and grafting, cuttings, bulbs, corms, rhizomes and tubers.

From the early 1970s, as a result of improved methods of *in vitro* cultivation and, new molecular biological procedures, plant cell and tissue culture has had a pronounced impact on both plant breeding and vegetative propagation (micropropagation) and has also simplified the storage and conservation of germplasm. In addition, plant tissue culture has been an integral part of the successful development of plant transformation technology, which has allowed the creation of transgenic plants by introducing DNA from almost any source.

The application of these procedures, known collectively as plant biotechnology, has become a versatile platform for basic scientific research as well as for commercial applications in agriculture, horticulture and forestry. This chapter will consider some of these applications and the following chapter will deal specifically with genetic modification.

MICROPROPAGATION

Micropropagation is a tissue culture technique for producing large numbers of identical copies of a plant from a tissue fragment of a 'mother plant'. In the simplest case, meristematic tissue from mature plants is removed under aseptic conditions and placed in or on sterile medium containing minerals, essential vitamins and an appropriate combination of plant hormones. The formation of the desired organs or tissues is induced by the action of the plant hormones. Usually, shoot development is induced on a medium designed for that purpose and the shoots are removed, separated and placed on a rooting medium, where roots and complete plantlets will be formed. These plantlets are genetically identical, and the result is therefore clonal propagation.

Micropropagation has the immense advantage of rapidly generating a large number of genetically identical plants in a much shorter time than could be achieved by conventional propagation methods. It has particularly important applications for genetic modification, where a novel transgenic plant can be rapidly multiplied.

Micropropagation has been applied successfully to the production of more than 1000 plant species. Many of these are ornamental, but the technique is also routinely applied to food crops and trees. It is particularly useful in several circumstances. The first is where it is troublesome to propagate a species vegetatively or where seed is recalcitrant (difficult to germinate). This includes ornamental species like the orchids. The second is where plant breeding is slow and introduces unwanted genetic variation, for instance in the case of tree species. The third is where the yield of a species is limited by a systemic disease that is transmitted by conventional propagation.

Commercial Micropropagation

Commercial micropropagation of plants is used for the production of large numbers of copies of a selected genotype. As only very small pieces of the parent plant need be removed, the plant is not destroyed and so rare or unusual plants can be used as starting material. Tissue explants, which may be the shoot tip, lateral buds, leaf, root or stem, are placed on media that encourage large numbers of shoots to differentiate. These shoots are then separated and placed on rooting media to form plantlets. By repeated subculture of buds or shoots many plants can be produced *in vitro* all having the same genetic characteristics as the original plant. The plantlets must then be weaned from the axenic conditions in which they were formed into viable plants capable of survival in conventional horticultural or agricultural environments.

The operational steps in micropropagation are labour-intensive and repetitive and in some production systems gains in speed; sterility and labour cost have been achieved by the introduction of automation. The relatively high production costs of *in vitro* micropropagation compared with classical methods of propagation are compensated for by the following benefits:

1. Rapid propagation because of short propagation cycles;
2. Large volume propagation of high-value ornamentals like orchids, and of trees and shrubs;
3. Storage and transport of large numbers of plants;
4. Production of plants is independent of season;
5. Supply can react quickly to changes in demand.

Micropropagation techniques are widely used to generate clones of high-yielding or disease-resistant individuals. Some species are genetically variable and while this may have advantages in conferring a range of disease resistance, it limits productivity in commercial production. The problem is heightened where productivity can only be determined after a long period, for instance a tree or shrub that does not begin to yield a commercial crop for several years. Many trees (both gymnosperms and angiosperms) and plantation crops like bananas (*Musa* spp.) and oil palm (*Elaeis guineensis*) show high levels of genetic variation and have been particularly suitable for the clonal propagation of plants selected for their agronomic traits (superior plants).

One of the first cultivated species to be improved in this way was the oil palm. Oil palms, which are grown extensively in Africa, Malaysia and Indonesia, account for 20% of world oilseed production. Yield can vary by as much as 30% per tree and 30 million trees are replaced annually. Conventional methods of vegetative propagation (cuttings, shoots or grafts) cannot be used for this plant with its obligate cross-pollination and whose natural means of propagation is by seed. Because of the substantial heterogeneity of plants grown from seed, desirable agronomic traits can only be fixed by clonal selection and propagation. Large-scale clonal propagation of selected plants is achieved by regeneration using somatic embryogenesis. The oil palm propagation programme has concentrated on yield improvement, ease of mechanical harvesting and disease resistance.

Similar principles of selection are used for the micropropagation of many other crops. Members of the plantain and banana families, for instance, are major crops of the humid tropics with an annual production of more than 85 million tonnes. Of this, around 10% is for export, the remainder being a staple for indigenous populations. *Musa* crops present a range of problems for improvement by conventional breeding methods. They are generally triploid and therefore sterile; they take 2 years from seed to seed and field trials require large amounts of space. Micropropagation of *Musa* based on shoot tips allows production of large numbers of pest- and disease-free plants. Together with other methods, it has permitted the production of faster-growing and higher-yielding plants that are popular with producers.

Other species commercially produced by micropropagation include: orchids, other flowers and ornamental plants, date (*Phoenix dactilyfera*), soft coconuts (*Cocos nucifera*), cardamom (*Elettaria cardamomum*); eucalyptus (*Eucalyptus globulus*), Chinese fir (*Cunninghamia lancolata*), rattan (*Calamus* spp.) and triploid water melons (*Citrullus lanatus*).

Production of Virus-free Plants

Plant viruses are systemic and transmitted by conventional propagation methods. The discovery that the meristem is often free of infection resulted in the first production of virus-free *Dahlia* in the 1950s. By the end of the decade, the technique had been used to remove paracrinkle virus from potato (*Solanum tuberosum*). The technique has subsequently been successfully applied to a wide range of crops and ornamental species. These include: banana, rhubarb (*Rheum raponticum*), strawberry (*Fragaria* spp.), citrus species, papaya (*Carica papaya*), apple (*Malus* spp.), sweet cherry (*Prunus aidum*), pear (*Pyrus communis*), grapes (*Vitis vinifera*), sweet potato (*lpomoea batatas*), *Allium* spp., taro (*Colocasia esculentus*), sugar cane (*Saccharum officinarum*) and ornamentals such as chrysanthemum (*Chrysanthemum coronarum*), carnation (*Dianthus caryophyllus*) and *Lilium* spp. It is likely that meristems remain virus free because they lack direct vascular connections to the rest of the plant and are continuously growing away from older, infected tissues. It may also be the case that virus movement is limited through plasmodesmata adjacent to the meristem.

To produce virus-free plants, virus-free meristematic tissue in shoot-tips or axillary buds of infected plants are cultured and regenerated to produce new plants. For success, the size of the explant, which varies with species, is the crucial factor. It must be as small as possible in order to eliminate all virus-infected tissue, but large enough for regeneration. If necessary, tissues may be heat-treated or incubated with an anti-viral agent like malachite green or thiouracil to kill remaining virus. A modification of this technique, known as micrografting, is used to produce virus-free woody fruit trees. In this procedure, sterile virus-free axillary buds are grafted onto a rootstock grown under sterile conditions. Potato and strawberry are examples of commercially grown crops that depend on a supply of virus-free stock each season.

Somatic Embryogenesis and Propagules

In the tissue culture techniques described so far, plantlets are frequently produced by meristem culture. As an alternative, embryos may be generated by cell and tissue culture (somatic embryos), either by the use of an embryogenic callus or from suspension cultures. Suspension cultures may be grown on a large scale and thousands of embryos generated. The embryos formed are genetically identical and can be used as if they were zygotic (seed) embryos. However the seed is a complex nutritional and protective system that greatly enhances embryo survival and facilitates propagation. Handling naked somatic embryos presents technical difficulties especially in commercial agriculture and horticulture where mechanized sowing of seeds is conventionally used. To overcome this problem, artificial propagules (or pseudo-seeds) have been developed, by surrounding the embryo with a hydrated gel. They have been produced for a variety of ornamental species, including lilies, azaleas and African violets, as well as for some trees and crops like tomato, asparagus (*Asparagus officinalis*) and celery (*Apium graveolens*). They can be successfully used for mechanical sowing.

Micropropagation based on somatic embryogenesis is also used worldwide for the commercial propagation of many major crops,, including rubber, sugar cane, potato and plantain. It is also widely applied in forestry and agroforestry. Trees like eucalyptus, poplar, bamboo, teak, spruce and pine are all commercially micropropagated in areas in which they are grown commercially.

Table 16.1. Forest trees micropropagated by the Canadian Forest Service

White spruce	*Picea glauca* (Moench) Voss
Black spruce	*Picea mariona* (Mill.) BSP
Red spruce	*Picea rubens* Sarg
Tamarack	*Larix laricina (Du* Roi) K. Koch
European larch	*Lorix decidua* Mill
Hybrid larch	*L. decidua Mill. X Lorix leptolepis* (Siebold & Zucc.) Gord.
Eastern white pine	*Pinus strobus* L.
Jack pine	*Pinus banksiana* Lamb.

Plant Breeding

Over the last 25 years, tissue culture methods in combination with other procedures have been applied to plant breeding. In many breeding programmes, it is now possible to use classical methods, new techniques or a combination of classical and new. In this section, we describe the tissue culture based methods and illustrate how their applications can be combined with other techniques.

In vitro Selection

The concept of *in vitro* selection is to exploit the genetic variation known to occur in plants by screening cell cultures for resistance to disease, insects, herbicides or stress. The procedure of *in vitro* selection typically involves subjecting cells in culture to a suitable selection pressure and recovering any variant cell lines that are resistant to the stress. These variant lines are then used to regenerate whole plants. The technique relies on a dependable method of regeneration from callus and presupposes that resistance displayed by undifferentiated cells in culture is equally expressed in whole plants and can be transferred to progeny by conventional plant breeding.

The use of *in vitro* screening to select for herbicide resistance is particularly effective as the herbicide can be added to the culture in defined concentrations. However, the mode of action of the herbicide is also an important factor, as it must be the same at cellular level as for the whole plant. Herbicide resistance has been selected by *in vitro* screening, for a range of herbicides in a number of plants. For example, glyphosate resistance in carrots was selected by subjecting carrot suspension cultures to repeated subcultures with incremental increases of glyphosate concentration in the media. Regenerated plants in the field were resistant to glyphosate.

In vitro screening for resistance to plant pathogens is most successful when the disease is caused by known *phytotoxic* compounds produced by the pathogen. Screening for resistance to the compounds or their analogues has led to the selection of plants resistant to a number of bacterial and fungal diseases, for example resistance to *Helminthosporium oryzae* in rice.

The frequency of genetic variation in plant populations can be increased by mutagenesis and mutagenized material can be subjected to *in vitro* screening to select desired variants. Mutagenesis can be brought about by:

1. Ionizing radiation;
2. Chemical compounds;
3. Cell and tissue culture regimes (somaclonal variation).

Mutagenesis induced by chemicals and ionizing radiation

There are several chemicals used to induce mutagenesis in the genomes of plants. The most used chemicals are alkylating agents, and the most preferred alkylating agent is the compound ethyl methane sulphonate (EMS). The, ionizing radiation may be neutrons, beta-radiation or X-rays. Seeds are mutagenized, germinated and progeny selected for the desired trait. Progeny may be rapidly micropropagated and backcrossed with other lines to combine the useful trait of the mutant with desirable existing features of the crop. About 2000 crop cultivars have been generated by mutagenesis and the FAO lists such cultivars available for crop breeding. Many current lines of crops, like rapeseed (*Brassica napus*), wheat (*Triticum* spp.), white bean (*Phaseolus vulgaris*) and barley (*Hordeum vulgare*), are the result of crosses with mutagenized lines. Thus, while mutagenesis has diminished as interest in plant genetic engineering has increased, it still remains an important technology.

Somoclonal variation

When an apparently genetically identical clone of plants, regenerated from cell and tissue cultures, is grown to maturity, some individuals show clear variation in characteristics—size, yield, colour etc. Some of these variants are useful; others are deleterious. These aberrants are known as *somaclonal variants* and may be used as part of a selection-breeding programme in crop improvement. Somaclonal variation leads to the creation of additional genetic variability. Somaclonal variation was used extensively in the 1980s and 1990s in plant breeding. The rate of variation exceeds that of natural mutation and results from the impact of the tissue and cell culture technique on the genome of plants being micropropagated. Most of the variations can be attributed to chromosomal instability and often the degree of instability is correlated with the length of time the cells were in culture. Since the early studies of Larkin and Scowcroft (1981) suggested that somaclones were a novel source of variation that could be exploited for crop improvement, the procedure has been applied to a wide variety of species. There are several examples where somaclonal variation was used to improve crop plants. These included melon (*Cucumis melo*), strawberry and rice (*Oryza sativa*) (disease resistance, food quality and salinity resistance, respectively), sugar cane (disease resistance), banana, wheat, barley, maize, tomato (disease resistance) and brassicas (development and maturation). Very few somaclonal variants are now being used commercially due to the genetic instability of the variants, and this strategy for crop improvement is now questioned. Somaclonal variation is also a major disadvantage when clonal uniformity is required, for example where tissue culture is employed for propagation of elite genotypes.

Somatic Hybridization

A novel and elegant use of cell and tissue culture technology for plant breeding is the production of somatic hybrids by protoplast fusion. Protoplasts are single plant cells that have had their cell walls removed by enzymatic digestion. Protoplasts can grow new cell walls to produce intact cells, each theoretically capable of regeneration to a whole plant.

The concept of somatic hybridization was based on the observation that when protoplasts are brought into close contact, sometimes they would fuse with each other. Subsequently, it was found that fusion could be greatly enhanced either by applying a brief electrical pulse or by treatment with compounds like polyethylene glycol (PEG).

The fusion of protoplasts from different species produce so called somatic hybrids. Somatic hybridization is an *in vitro* technique that makes it possible to circumvent the incompatibility systems that prevent *in vivo* hybridization between sexually incompatible species. The first successful fusion was between two species of *Nicotiana*, and reports of other fusions soon followed, including the somatic hybrid from tomato and potato.

Early expectations that somatic hybridization would play an important role in the development of new varieties have remained unfulfilled. However, some notable applications have resulted from somatic hybridization in closely related species. Fusion between *Nicotiana tabacum* and *Nicotiana rustica* has produced tobacco plants that varied in nicotine and tar contents, as well as their resistance to blue mould and black root rot. Somatic hybridization has been used extensively with citrus, for instance to produce interspecific hybrids between the two sexually incompatible species *Citrus reticulata* and *Citropsis gilletiana*. Hybrids have also been produced between apple species.

The fusion of haploid protoplasts has been used to shorten breeding programmes, and fusions between the nuclear genome of one species and the cytoplasmic genome (mitochondria and chloroplasts) of another have produced a fusion product called a *cybrid* (cytoplasmic hybrid). Cybrids are of particular interest in producing cytoplasmic male sterile (CMS) lines for hybrid breeding, as the CMS trait is carried by the mitochondria. CMS cybrids of rice have been produced and used in hybrid seed production and similar cybrids have been produced for other crops.

Haploidy

The production of haploid plants (plants having only one set [*n*] of chromosomes) is one of the major contributions of plant tissue culture to plant breeding. Haploid plants can be induced to double their chromosome number (2*n*) by chemical treatment, producing double haploids (dihaploids) that are homozygous for all genes. Dihaploids are desirable as parents for breeding of F_1 hybrids and facilitate the selection of recessive traits. Tissue culture techniques have simplified breeding programmes by dramatically reducing the time required to produce haploids. There are two commonly used *in vitro* methods for the production of haploids:

1. Anther culture;
2. The culture of microspores.

In both cases, embryogenesis is induced in the cultures and the embryos used to regenerate whole plants.

Haploids have been used in plant breeding to increase yield and improve traits like time to maturity and resistance to abiotic stress. There are examples of cultivars resulting from haploid breeding in many species; China has been particularly active in this area, with new varieties of rice, wheat, tobacco and hot and sweet peppers (*Capsicum* spp.). It was estimated that in 1995, 1.5 million ha, mostly of rice and wheat produced by haploid techniques, were under cultivation. Elsewhere, rice, Chinese cabbage (*Brassica rapa*), broccoli (*Brassica oleracea*), barley, maize, tobacco, strawberry and asparagus have been generated by haploid breeding:

Embryo Rescue and Culture

Embryo culture was one of the first successful plant tissue culture techniques attempted. In 1904, Hannig cultured mature embryos of the crucifers *Raphanus* and *Cochleria* using aseptic techniques. The principal application of embryo culture is to rescue embryos produced as the progeny from interspecific (wide) crosses that normally fail to develop viable seed. In these crosses, post-zygotic incompatibility mechanisms caused early abortion of embryos. This barrier to embryo development can be overcome by removing the immature embryo from the developing seed and culturing it *in vitro* to produce the hybrid plant. Wide crosses and embryo rescue are important tools for the transfer of desired traits such as disease resistance, stress tolerance or high yield from wild species to crop species.

Biodiversity and Conservation of Germplasm

The intensification of agricultural production seen during the second half of the 20th century with the emphasis on increasing output by the largescale cultivation of high-yielding elite clones, has reduced both the number and the genetic diversity of the species used in agriculture. Crop uniformity and neglect of the genetic diversity found in wild related species has had the effect of narrowing the gene pool from which parental material has been drawn. The result of the narrowing of the genetic base is an increased vulnerability of crops to pests, diseases and adverse abiotic factors. There is an increasing awareness that the genetic resources of wild and cultivated plants together constitute a wide pool of agronomically valuable genes that will be needed in the decades ahead, to increase tolerance of drought and poor-quality soils, boost resistance to diseases, maximize yields and improve nutritional quality. In response to this problem, there has been a worldwide effort to collect and conserve plant genetic resources and to develop techniques for the long-term storage and maintenance of germplasm.

The are three main categories of genetic resources of crop plants that are conserved:

1. Germplasm from the wild relatives of crop plants;
2. Germplasm derived from traditional breeding programmes;
3. Biotechnologically derived plant germplasm.

Although field gene banks have been an important means of conservation, seed storage has been the most widely used method of conserving germplasm. Conventionally dehydrated seeds are stored at low temperatures, in cold rooms. However, many species cannot be conserved by this method because their seeds do not tolerate dehydration and/or cold (e.g. coffee, *Coffea* spp.), or they are propagated vegetatively (e.g. banana). Plant tissue culture has provided an alternative system for both *in vitro* collecting and in vitro conservation of germplasm.

In vitro Collecting

Essentially, the technique of *in vitro* collecting involves taking an explant in the field, decontaminating the material and placing it in sterile culture medium before transfer to a plant tissue culture laboratory for further *in vitro* processing. This technique is particularly effective for species with recalcitrant seeds or embryos that deteriorate rapidly.

In vitro Culture and Germplasm Storage

In vitro culture offers the advantage that disease-free cultures can be established from explants and stored under optimal conditions for long periods. For instance, virus infection can be eliminated by meristem culture as described earlier. In addition, germplasm stored as disease-free cultures can be easily transferred between countries without the need for quarantine restrictions.

In vitro culture collections may consist of dedifferentiated cell cultures such as suspension and callus culture, or organized tissue cultures like embryos or meristems. Cultures are maintained either in slow-growth storage or by cryopreservation.

Slow-growth storage reduces the frequency of subcultures and saves on labour and media costs. Slow growth is achieved by a variety of methods including low temperature (2-8°C), low light, low oxygen, desiccation or culture on minimal medium with growth retardants.

Cryopreservation is the storage of cultures at or near the temperature of liquid nitrogen (–196°C). At this temperature, all cellular processes are effectively stopped and germplasm can be stored for very long periods without risk of contamination. Cryopreservation is achieved by a number of different procedures that either use cryoprotectants and controlled cooling or apply dehydration followed by rapid freezing.

The techniques of *in vitro* storage go beyond that of preserving germplasm for plant breeding; they also offer possibilities of maintaining material of rare and endangered species.

17

Tissue Culture in Genetic Engineering

As described earlier, plant tissue culture is also a prerequisite for the genetic engineering of transgenic plants with novel characteristics. The technical operation of introducing and expressing foreign genes in plants was first described for tobacco in 1984. Since then this technology has been extended to over 120 species and large areas of transgenic crops are now cultivated worldwide.

For the greater part of the 20th century, plant breeding was the traditional method of generating cultivated plants with desired characteristics. Selective plant breeding (hybridization) brings together desired genes from two or more individuals with a resultant combination of desired traits in the offspring—a hybrid. Selective plant breeding has combined applied Mendelian genetics with other methods, such as the induction of mutations, to achieve a number of specific breeding aims. For example, crop yields have been increased, qualitative traits, such as palatability and aesthetic features, have been improved and resistance to disease and pests introduced.

The creation of new genotypes by classical plant breeding largely depends on mechanisms that occur during meiosis in germ cells. In contrast, modern biotechnological methods bypass the generative cycle and produce new genotypes by introducing desired ones directly into somatic cells. Furthermore, the development of tissue culture and molecular biological procedures for the exchange of DNA between unrelated organisms has enabled novel genes derived from organisms outside the plant kingdom to be introduced into recipient plants. In addition, this modification of plant genomes by the integration and expression of foreign DNA (genetic engineering) leads to the formation of new genotypes (transgenic plants) that can be developed further by classical plant breeding methods.

Genetic Engineering

Four key steps are required to produce a transgenic plant.

1. Identification and isolation of the gene(s) of interest from the original organism.
2. Integration of the isolated gene into a plasmid and multiplication in bacteria (cloning).
3. Insertion of the cloned gene into cells of a target plant.
4. Selection and growth of transformed cells into new plants.

Each step requires specific tools and procedures.

Identification and Isolation of the Desired Gene

The essential first step in genetic engineering is finding and making copies of the specific gene that encodes the protein of interest. A range of techniques is available for locating and sequencing stretches of DNA that contain a desired gene. Commonly, these procedures include identification of the desired gene through screening of a DNA (gene) library of the donor organism. Screening is based on detecting the DNA sequence of the cloned gene, detecting the protein that the gene encodes or by using linked DNA markers. The identified gene is joined by ligation into a bacterial plasmid (a cloning vector) that is copied (cloned) by DNA replication in the host bacterium. Frequently, the gene of interest is cloned from a genomic DNA library and subcloned from a cDNA library. The cDNA form of the gene is preferred to the genomic DNA form, as cDNA transcripts will require no RNA processing before translation. This is particularly important in heterologous expression, as the host plant may not have the appropriate machinery to produce functional messenger transcripts.

Construction of Transgene and Integration in Cloning Vector

The isolated gene is integrated into a bacterial plasmid (a cloning vector) and multiplied in a host bacterium like *Escherichia coli*. The cloned gene must be modified so that it will be effectively expressed when inserted into the plant. Typically the modifications involve changes to sequences in the region of the gene that control gene expression, however sometimes the cloned gene itself is modified to achieve greater expression in a plant. For example, changes can be made to overcome codon bias. Plants prefer G-C nucleotide pairs compared to bacterial genes, which have a high percentage of A-T nucleotides. In cloned bacterial genes, A-T nucleotides can be substituted for G-C nucleotides without significantly changing the amino acid sequence, but enhancing the production of the gene product.

Promoters and gene expression

In addition to suitable start and termination sequences, each construct must contain an appropriate promoter. The promoter ensures that the protein encoded by the gene is produced either constitutively, during a particular phase of development or in specific organs. The most commonly used promoter in plant transformation is the 35S promoter (35S CaMV). The 35S CaMV promoter is a constitutive promoter derived from the DNA-based cauliflower mosaic virus (CaMV) and is used to regulate the expression of transgenes in several transgenic crops currently under cultivation. The 35S CaMV promoter can function with both monocots and dicots and maintains a high and constant level of transcription.

In many genetic engineering experiments, tissue or developmental-specific expression is an important requirement. Consequently, a great effort is currently being made to identify plant promoters of known specificity. A tissue-specific plant promoter that is currently used in transgenic crops is the phosphoenolpyruvate (PEP) carboxylase promoter. The PEP carboxylase promoter is from the plant gene encoding a photosynthetic enzyme and this promoter will be active only in cells that are producing photosynthetic proteins. Expression also slows and stops at the end of the growing season when the requirement for photosynthesis has diminished.

Selectable markers and reporter genes

In addition to the desired gene, transformation vectors are constructed with a second gene, either a selectable marker gene or a reporter gene. These genes have no designated function either in the transformed plant or in any products derived from it, but are necessary to identify cells or tissues that have successfully integrated the transgene. This is essential because the frequency of achieving incorporation and expression of transgenes in plant cells is generally limited to a few per cent of the targeted tissues or cells. Selectable marker genes and reporter genes are inserted in the same stretch of DNA as the gene of interest where they are fused to the same promoter and they require termination sequences for proper function.

Selectable marker genes encode proteins that provide resistance to agents that are normally toxic to plants. Only those cells that have

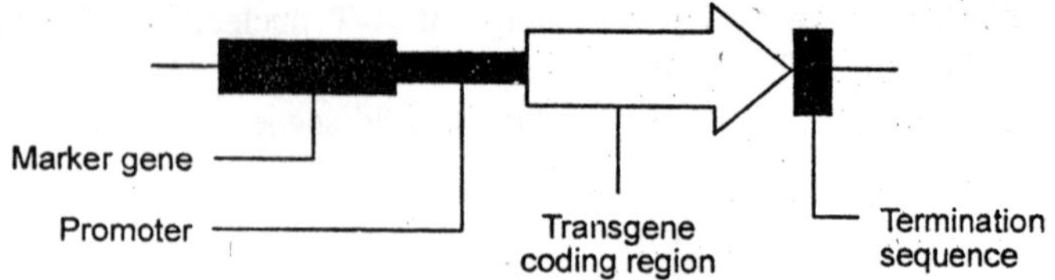

Fig. 17.1. A simplified scheme showing the regions (and their relative positions) of transgene construct that are required for integration and expression.

integrated and expressed the selectable marker gene will survive when grown on a medium containing the appropriate toxic compound. In contrast, reporter genes do not affect survival on toxic media but code for gene products that have easily detectable phenotypes. Selectable marker genes are almost exclusively of two types: (i) genes conferring antibiotic resistance and (ii) genes conferring herbicide resistance.

There are several commonly used antibiotic resistance markers. Currently, the most widely used antibiotic resistance marker is the neomycin phosphotransferase type II gene (*nptII*), which is derived from the bacterial transposon Tn5. Neomycin phosphotransferase II (NPTII) is an enzyme that confers resistance by inactivating a number of related antibiotic aminoglycosides, such as kanamycin, paromomycin and geneticin. Transformed cells expressing NPTII are protected from the effects of the kanamycin and can grow and regenerate into whole transgenic plants in cell culture media containing the antibiotic. A number of other antibiotic resistance markers are employed in the production of transgenic plants. These include the *aad* gene for the enzyme 3'(9)-*O*-aminoglycoside-adenylyltransferase, that confers resistance to streptomycin and spectinomycin and the *hpt* or *hph* gene from *E. coli*, which confers resistance to hygromycin. The resistance gene codes for a kinase that inactivates hygromycin through phosphorylation.

Two herbicide selectable markers are commonly used in plant transformation. The first is the *bar* gene encoding the enzyme phosphinothricin acetyltransferase (PAT), isolated from *Streptomyces*. PAT confers resistance to the herbicide phosphinothricin by catalysing its detoxification to the acetylated form. The second herbicide marker gene is the mutant allele of the aroA locus of *Salmonella typhimurium* encoding a 5-enolpyruvylshikimate-3-phosphate (EPSP) synthase in which a single substitution of a proline for a serine causes a decreased affinity for the herbicide glyphosate (Roundup), without affecting the kinetic efficiency of the enzyme. Non-transformed cells can survive but have an altered phenotype.

Reporter genes

After transformation, there are several ways of determining where and when the gene of interest is expressed in the host plant tissues. Some of these procedures, such as Southern or northern hybridization, dot blot analysis, enzymatic assays or polymerase chain reaction (PCR) are either laborious or require careful optimization. Reporter genes are a rapid and convenient strategy to characterize the expression patterns of the transgene. With most reporter genes, successful transgene expression can be determined through a visual assay.

Often, the reporter gene encodes a protein that can be easily detected either by characteristics such as fluorescence (fluorogenic) or through an enzyme activity that produces a coloured product (chromogenic). Thus, the location and amount of gene expression in a transformed tissue can be readily assessed.

The β-galactosidase (lacZ) and β-glucuronidase (GUS) genes are two examples of reporter genes that are frequently used. An important requirement of reporter genes is that their activity is absent in the plant in which they will be used. Both *lac*Z and GUS are genes derived from *E. coli*; however, some plants do contain some β-galactosidase activity and this can lead to high background staining. GUS activity is normally very low in plants, and so it is a common reporter gene used in plant studies. GUS activity is visualized by transferring the transformed tissue to a medium containing 5-bromo-4-chloro-3-indoyl-l-glucuronide, which is a substrate for the β-glucuronidase enzyme. Cleavage of the substrate by β-glucuronidase produces a blue-coloured product. The expression pattern determined by the reporter gene can then be confirmed by one of the other more quantitative strategies like Southern analysis.

Genetic Transformation

Transformation is the insertion and integration of the DNA representing the cloned gene into the genome of the host plant cell so that it expresses the protein encoded by the gene. It is important that the foreign DNA is integrated into the host plant chromosomes so that it can be passed on during mitosis to future cell generations. This is known as '*stable transformation*' and transformed cells can be used to regenerate a complete plant where every cell contains a copy of the foreign gene. If the DNA of the inserted gene is not integrated into a chromosome, it can still be expressed for a short time, a phenomenon known as '*transient expression*'.

Table 17.1. Activities and measurement of proteins encoded by reporter genes used in plant transformation

Protein	*Activity*	*Measurement*
GUS (β-D-glucuronidase)	Hydrolysis of β-D-glucuronides	(i) With the chromogenic substrate 5-bromo-4-chloro-3-indolyl-β-D-glucuronide, the product formed (indoxyl derivative) is oxidized to a blue precipitate. (ii) With the fluorogenic substrates 4-methylumbelliferyl-β-D-glucuronide (MUG). Fluorescence can be measured at an excitation of 360 nm and an emission of 450 nm.
LacZ (β-D-galaetosidase)	Hydrolysis of β-D-galactosides	(i) With the chromogenic substrate 2-nitrophenyl β-D-galactopyranoside. The product is 2-nitrophenol (yellow in alkaline solution) (ii) With fluorogenic substrate 3-carboxyumbelliferyl β-D-galactopyranoside (CUG).The fluorescent product is measured at an excitation of 390 nm and an emission of 460 nm.
LUC (luciferase)	Oxidation of luciferin	Luminescence detection (photon emission)
GFP (green fluorescent protein)	Fluorescence	Irradiation with blue light

The insertion of genes into plant cells can be achieved by the following two types of procedures:

1. Indirect gene transfer that exploits the naturally occurring vector *Agrobacterium tumefaciens*.
2. Direct gene transfer, which is based on the physical methods that drive the uptake of DNA.

Indirect gene transfer

A. tumefaciens is a Gram-negative soil bacterium that infects susceptible plants causing the formation of tumorous growths that are characteristic of a pathological condition known as crown gall disease. The bacterium attaches to wounded tissue at the base of the stem and induces multiplication of the underlying cortical cells. During the process of infection, an independent loop of *Agrobacterium* DNA (more than 200 kb) known as the 'Tumour-inducing' or Ti-plasmid is inserted into the host cell. A mobile DNA segment of the Ti-plasmid (known as *T-DNA*) is transferred to the nucleus of infected cells where it integrates into the host's genome and is transcribed like a normal part of the plant's DNA.

T-DNA contains two types of genes: oncogenic genes encoding the enzymes responsible for the biosynthesis of phytohormones (auxins and cytokinins) that are responsible for tumour formation; and genes encoding proteins involved in the synthesis of a group of compounds known as opines that are utilized by the bacteria as sources of nitrogen and carbon. The T-DNA is flanked by 25-bp imperfect repeats that define the boundaries of the T-DNA and act as a *cis* element signal in the transfer and integration process. T-DNA transfer is mediated by the combined action of genes encoded by the virulence region (35 *vir* genes) of the Ti-plasmid. Wounded cells of host plants secrete low molecular weight phenolic compounds (acetosyringone and hydroxyacetosyringone) that stimulate the *vir* genes.

Agrobacterium-mediated transformation exploits this natural transformation mechanism. The genes within the T-DNA region are not required for DNA transfer and can be removed and replaced by desirable genes making it possible to use the Ti-plasmid as a transformation vector without disease induction. Transformation vectors and *Agrobacterium* host strains that are no longer oncogenic (i.e. they are disarmed) have been developed for plant transformation.

Transformation vectors

A primary requirement for *Agrobacterium*-mediated transformation is inserting the gene of interest into the T-DNA in preparation for transfer to the plant cell. Two classes of transformation vectors have been developed to achieve this aim.

1. Integrating (or cointegrate) vectors, where the gene to be transferred is engineered into T-DNA on a disarmed Ti-plasmid containing *vir* genes.
2. Binary vectors, where T-DNA modified to carry the desired gene and the *vir* region reside on separate plasmids.

Binary vectors are now the most commonly used strategy as they are easier to manipulate.

Agrobacterium transformation is a commonly used method of transformation for dicots because of the flexibility and ease of the procedure. In addition, single-copy integrations are the most common event, thus reducing potential problems of co-suppression and instability of the transgene.

Plant transformation is achieved by incubating (co-cultivating) *Agrobacterium* with a suitable explant. The bacterium is killed with an antibiotic and transformed cells are selected on an appropriate screening medium. Selected cells are allowed to proliferate in tissue culture and used to regenerate transgenic plants. Explants that are typically used for transformation in dicots include leaf discs, stem segments, cotyledons, callus suspension cultures, protoplasts and germinating seeds.

In some species, a number of factors are found to be crucial for successful transformation. These include the *Agrobacterium* strain and the temperature and duration of co-cultivation. *Agrobacteria* are attracted to wounded plant cells and pre-wounding of tissue with glass beads, microprojectiles, sonication or the addition of hydroxyacetosyringone has been used to increase transformation efficiency in some species.

It was previously thought that monocots were recalcitrant to *Agrobacterium*-mediated gene transfer because of a barrier to T-DNA integration. However, recently reliable and efficient procedures have been established for rice, maize, wheat, barley, banana and sugar cane.

Direct gene transfer

A major obstacle for all gene transfer systems including biological delivery systems is the impenetrability of the plant cell wall. Direct gene transfer techniques attempt to overcome this problem by using physical methods to facilitate the uptake of naked DNA by plant tissues.

Gene gun

The most frequently used method of direct gene delivery is termed *biolistic* or *ballistic* transformation. This technique uses the mechanical force generated by high-velocity particles (microprojectiles) to propel DNA through the biological barriers of plant cells. In essence, particles of gold or tungsten (0.2-0.3 μm) are coated with DNA and shot into target cells by acceleration.. The acceleration is generated by a particle gun using gunpowder, high-pressure gases (helium, carbon dioxide or nitrogen) or an electric discharge. This technique has the potential to deliver DNA into any tissue from any species of monocots and dicots, and to all genomes in the plant cell including those of the mitochondrion and chloroplast. In some instances, pre-treatment of the tissue before bombardment has improved transformation. For example, proper pre-

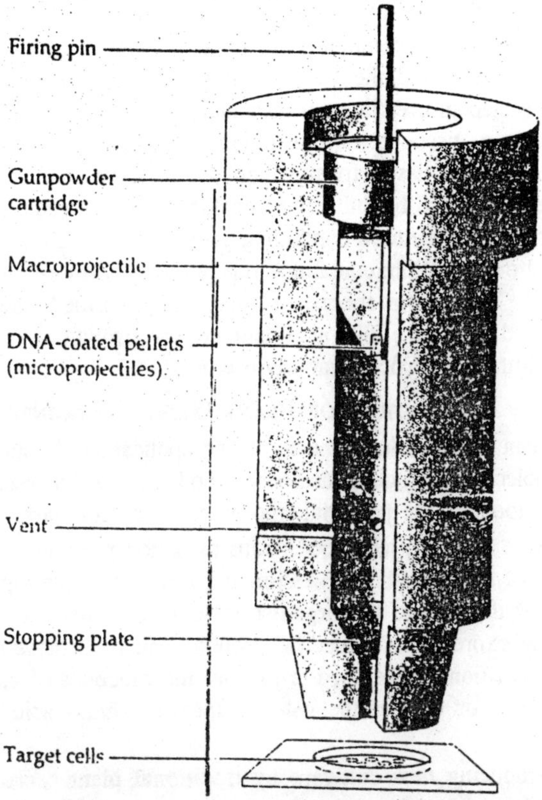

Fig. 17.2. Inserting DNA into cells. This "DNA particle gun" — shoots DNA into eukaryotic cells.

culturing of the explant tissue by an osmotic adjustment delivered by partial drying or by the addition of an osmotic agent to the medium.

Alternative methods

Several alternative methods of direct gene transfer have been devised but for practical reasons they are not often used. Polyethylene glycol (PEG), an agent known to induce fusion of biological membranes has been used to promote the uptake of isolated DNA into plant protoplasts. Other ways of facilitating the uptake of naked DNA by plant cells include electroporation, laser microbeams, silicone carbide whiskers and microinjection with small cannulas.

Recovery and Regeneration

Irrespective of the transformation system used, successful development of a transgenic plant depends critically on the ability of the transformed tissue to produce totipotent cells that proliferate and regenerate into a complete viable plant. The selection of successful transformants, and their regeneration and propagation, is the rate-limiting step in creating a transgenic plant and these steps depend on a variety of tissue culture techniques. Callus is the most commonly used totipotent tissue from which whole plants can be regenerated. Callus is initiated by culturing transformed tissue such as leaf discs in a nutrient-rich medium containing growth hormones, and plantlets are formed on a medium containing regeneration hormones (cytokinins and auxins). When both shoots and roots have formed, the plantlets are separated and carefully acclimatized for growth in a conventional agricultural or horticultural environment.

Applications of Plant Genetic Engineering

Plant biotechnology (the combined application of plant tissue culture and molecular biological methods) is both a powerful research technique acid a tool for the potential improvement of cultivars.

As a research technique, transformation provides an unparalleled means of studying the expression and function of plant genes from the level of the cell to the organism. Novel proteins can be engineered and the expression of specific proteins can be removed. In addition, transformation can be used to explore the functions of various sections of a gene or dissect the role of specific amino acid residues in a protein.

Producing new cultivars by traditional plant breeding is a slow and labour-intensive process. It takes several years to develop an improved variety since several crosses must be made, seeds collected and germinated and the resulting plants grown before the results can be assessed. An additional and significant limitation of traditional plant breeding is that it can only use genes from within one species or closely related species for cultivar improvement. Furthermore, breeding typically involves transfers of whole sets of inherited characteristics and any undesirable traits introduced during initial crosses must later be meticulously bred out. Biotechnological procedures on the other hand offer the possibility of using genes from other kingdoms and species, improving the precision of transfer of desirable traits and generally speeding up the production of new cultivars.

The greater part of the current research in plant biotechnology focuses on food crops (80%); the remainder on non-food crops like cotton (*Gossypium* spp.), tobacco, ornamentals and on pharmaceutical production. Several crop plants have been genetically modified and some of these have been approved for commercial cultivation in some countries. Examples of genetically engineered crops that are currently grown commercially are described below

Herbicide-resistant Crops

In modern agriculture, weed-control is effected by selective herbicides that can suppress or kill yield-reducing weeds without harming the crop. The most common factors responsible for selectivity are: (i) the ability of the crop to metabolize the herbicide faster than the competing weed and (ii) the sensitivity of the protein targeted by the herbicide in the weed and the insensitivity of the corresponding protein in the crop plant. Several crops have been genetically modified to be specifically resistant to some herbicides (eg. glyphosate,: glufosinate and bromoxynil) that are widely used. Glyphosate (Roundup, N-(phosphonomethyl) glycine) is a nonselective herbicide, which means it is active against most plants. Glyphosate is not metabolized by plants and specifically inhibits the enzyme 5-enolpyruvylshikimate-3-phosphate synthase (EPSPS) that catalyses a critical step in the synthesis of aromatic amino acids. Crop plants have been engineered to be insensitive to glyphosate either by transferring a bacterial gene for EPSPS that is insensitive to glyphosate, thus providing a bypass of the plant's EPSPS, and/or a bacterial gene that encodes a protein that catalyses the metabolism of the herbicide.

Insect-resistant Crops

Bacillus thuringiensis is a soil bacterium that produces insecticidal proteins known as the crystal proteins, Cry proteins or Bt toxins. There are different forms of protein produced by different subspecies of the bacterium, each of which is specifically active against one or two insect orders. When Bt protein is ingested by larvae of the target species, the toxin binds to cells of the gut causing lysis, paralysis and eventually death. Several crop plants, notably maize and cotton, have been transformed with various genes for Bt toxins to produce transgenic plants that are protected from insect attack and consequent yield losses.

Controlled Ripening of Fruit

Transgenic tomatoes have been engineered with delayed ripening of the tomato fruit. In one example, the genetic modification causes a down-regulation of the gene encoding polygalacturonase. Polygalacturonase is an enzyme that catalyses the cleavage of pectin chains. In the cell wall of the fruit during ripening and which results in softening. By down-regulating polygalacturonase, tomato fruit soften more slowly during ripening than conventionally bred tomatoes. This allows harvesting to be delayed until the fruit has fully ripened, improving flavour and conferring improved processing properties.

Improving the Nutritional Value of Food

Rice is a staple food in many parts of the world. The edible part of the rice grain, the endosperm, lacks provitamin A (β-carotene), which is converted in the human body to vitamin A. By engineering rice plants with a combination of three transgenes, the metabolic pathway for the biosynthesis of provitamin A in the endosperm was achieved. This provitamin rice known as Golden rice is still being developed.

18

Culture of Industrially Important Plant Cells

It is not more than 100 years ago that chemical synthesis started to replace plants as the major source of organic compounds for human use. Today, plant serve as an important source of secondary metabolites used in pharmacy, biotechnology and food technology. The first practical applications of *in vitro* techniques to plants occurred in the 1920s, with differentiated structures such as zygotic embryos. Efforts to grow dedifferentiated, isolated plant cells *in vitro* were unsuccessful until 1939 and it was the discovery of auxin in 1930 that finally led to success in achieving the production of continuously growing plant cell cultures.

Soon after the Second World War, many examples of interesting biosynthetic processes in dedifferentiated plant cells became known. In some plant cell cultures even higher concentrations of secondary metabolites were found as compared to the intact plants. At the same time, cell culture techniques for the cultivation of suspension cultures were established and improved. It may have been the commercial production of antibiotics by fungi that led plant physiologists to think that the production of secondary plant metabolites by dedifferentiated cell cultures was a possibility. Although, for a long time, the idea to produce commercially important secondary metabolites by dedifferentiated suspension cultures remained the major concern of plant biotechnology, up to now, there have been very few examples of commercial applications. This situation is due to a number of limitations. Unfortunately, many compounds of high commercial value (e.g. morphine, codeine or cardiac glycosides) are not formed by dedifferentiated cell lines. In other cases, compounds are only formed in small amounts, such that commercial exploitation seems to be impossible. Different strategies have been developed to overcome these problems such as the enhancement of production rate by the alternating use of growth and production media, the application of biotic and abiotic elicitors, and the selection of high yielding strains.

The basis for the selection of high yielding strains is the heterogeneity that occurs in almost all cell lines. Chromosomal aberrations have been observed for single cells of many plant cell cultures. More important for biotechnology is the fact that cells can also differ largely in their production capacity for certain compounds. In most cases it has not been clearly demonstrated whether this heterogeneity has a genetic or an epigenetic basis. Unfortunately, it is often the case that selected, high yielding strains show considerable instability when a selection pressure is no longer maintained. It is obvious that the stable formation of a desired compound is an essential prerequisite for any commercial biotechnological application, especially for the capital-intensive large scale production of plant metabolites. Therefore, the major concern of cryopreservation of cell cultures for biotechnological applications is not just survival (sometimes not even genetic stability), but mainly the stability of biosynthetic capacity. Genetic changes concerning genes that are not expressed in culture may not be of importance; on the other hand irreversible epigenetic changes, that is, changes in gene expression are crucial for the maintenance of biosynthetic capacity. Since the biosynthetic capacity may vary from cell to cell, even in selected high yielding strains, the major risk for conservation using cryopreservation is an undesired selection process, which might occur if the survival rates vary from cell type to cell type.

The following review will therefore concentrate on studies in which cryopreservation has been applied for the conservation of secondary product forming cells. Emphasis will be given to investigations which assess the stability of product formation after the recovery of cells from cryopreservation.

Stability of Product Formation after Cryopreservation

Anthocyanins

The first study on the retention of product formation after cryopreservation was performed by Dougall and Whitten (1980). These authors investigated the formation of anthocyanins in 25 sub-lines of a wild carrot cell culture before and after storage at −140°C. Anthocyanins are normally formed spontaneously in *Daucus carota* cell lines and they are not of great pharmaceutical interest, but some are used in food technology. The low costs of existing production methods would not normally justify the production of these secondary metabolites by cell culture technology.

Dougall and Whitten (1980) used a single, simple freezing method for all 25 different sub-lines. Thus, without a pre-culture phase, cell density was adjusted to a specific value; medium with dimethyl sulphoxide (DMSO) was added to a final concentration of 5 per cent. A freezing rate of −1 °C/min was used to a final temperature of −70°C at which the samples were then immersed into liquid nitrogen. Unfortunately, no measurements of survival rates were reported by these workers, but they state that from two samples per cell line, at least one re-grew. For detecting the total content of anthocyanins, the absorbance of extracts at OD_{530} was measured and Dougall and Whitten (1980) compared product formation in a maintenance and a production medium before and after cryopreservation. Before freezing, the high accumulating cell lines showed less growth than in the production medium. Although growth rate after cryogenic storage was decreased, especially for the low yielding cell lines, the capacity for product formation remained the same for the different cell lines. Thus, 'high yielding' strains remained high yielding and low yielding strains remained 'low yielding'.

Ginsenosides

One of the early targets for the commercial application of cell culture technology was *Panax ginseng*, the root extracts of which are traditionally used in Asian medicine and in increasing amounts in the Western hemisphere. Conventional plant production is time consuming and costly. The plants need a cultivation period of five to six years and harvesting the roots normally destroys the plants. Fortunately, cell cultures spontaneously produce almost the same pattern of ginsenosides as the whole roots of plants and the contents of ginsenosides can quite often exceed that of roots. Work on *Panax ginseng* cell cultures has also led to cryopreservation studies.

The first paper on the subject was published by Butenko et al. (1984). They used a programmable freezing approach. For pre-culture, they combined osmotic treatments with cold hardening. A comparison of different methods revealed that the best result was achieved using a pre-culture procedure which decreased the cultivation temperature gradually from an ambient temperature to 2°C and, simultaneously increasing the sucrose content of the medium from 3 per cent to 20 per cent within 18 days. For cryoprotection, 20 per cent sucrose yielded the highest survival rates (which was 51 per cent, measured by phenosafranine staining). The rather complicated cooling programme included a seeding (ice nucleation) step and the cells were transferred into liquid nitrogen from -70°C. In terms of post-cryopreservation stability assessments, only the growth curve of control and recovered cultures were compared one years after recovery; they appeared unchanged. No data on product formation were published.

Detailed data on product formation of recovered cultures were later produced by Seitz and Reinhard (1987). They used different approaches, all based on programmed freezing and they also used the slightly modified procedure of Butenko et al. (1984). Butenko's method based on cold hardening proved a superior approach to using mannitol or sorbitol as pre-culture treatments. Using a more simple freezing programme, Seitz and Reinhard (1987) even achieved survival rates of up to 40 per cent. With four different methods they were also able to recover actively growing cell cultures and all recovered cell lines showed the same growth characteristics as the controls. For assessing chemical properties, Seitz and Reinhard (1987) measured not only the contents of total saponins but compared the production patterns of eight ginsenosides. The interesting result was that those which yielded much lower survival rates, had recovered cell lines with an unchanged product pattern. It was also notable that one of the methods showing less than 10 per cent survival showed better recovery growth in the survivors, than two other methods showing much higher survival rates.

Mannonen et al. (1990) preserved a *Panax ginseng* cell line using the standard procedure of Withers and King (1980). Unfortunately, no post-freeze survival rates were given; however, the total amount of ginsenosides, together with product pattern was unchanged after cryopreservation. Furthermore, these authors demonstrated that alternative conservation methods, such as preservation under mineral oil for six months as well as continuous subculturing for 14 months failed to preserve the biosynthetic capacity of the cells.

Diosgenine

Diosgenine has been a compound of extraordinary pharmaceutical importance for many years. It was used as raw material for the semi-synthetic production of steroidal pharmaceuticals and until 1975, the main product source was *Dioscorea* roots collected in Yucatan, that is, until the plant almost became extinct. Thus, parallel efforts were made to find chemical structures for synthesis, and alternative plant sources were explored. The use of cultured cells of *Dioscorea* to produce diosgenine was investigated and cryopreservation became an important means of storing high producing cell lines. Thus, Butenko et al. (1984) published a successful cryopreservation method for *Dioscorea deltoidea* cell cultures. They used a programmable freezing method and for pre-culture, amino acids in low concentrations (0.01-0.02 M) were added. Asparagine and alanine yielded the best results; proline was less effective. DMSO 7 per cent was used as cryoprotectant and programmed freezing down to -90°C before immersion in liquid nitrogen was applied. Extracts of recovered and control cultures were analyzed for their saponin and phytosterol content by GC. Analysis was performed in the second and fifth passage after thawing and the authors quantified diosgenine, sitosterol and stigmasterol. The secondary product content of samples recovered from liquid nitrogen were equal to unfrozen controls. Even more impressive was the fact that the GC scans of recovered cryopreserved cells and control cultures had identical profiles, in terms of product pattern and the magnitude of their side peaks.

Rosmarinic Acid

The caffeic acid derivative, rosmarinic acid is formed spontaneously and in high amounts by cell cultures of several plants. It gained commercial interest because of its antiphlogistic activity and rosmarinic acid was produced even under large scale conditions. For a long time the contents of rosmarinic acid in *Coleus blumei* cell cultures was the highest of all secondary metabolites measured in plant cell cultures.

Cryopreservation of dedifferentiated cell cultures of *Coleus blumei* was carried out in Tubingen with an extensive study on the formation of rosmarinic acid performed by Reuff (1987). The optimized freezing method is very simple: cells from the logarithmic growth phase are incubated for eight hours in medium containing 1 M sorbitol, cooled at -1°C/min to -40°C, held at this temperature for 40 minutes and then immersed into liquid nitrogen. Analyses of recovered cultures showed that the production rate of rosmarinic acid, as measured by HPLC, was not changed by cryogenic storage. Even more interesting, is that Reuff (1987) analyzed the production rate of rosmarinic acid during several steps of the optimization procedure. Thus, changing the transfer temperature to liquid nitrogen from -100°C to -40°C, for example increased the survival rate from almost 0 to 25 per cent. Even cultures recovered from experiments using a sub-optimum transfer temperature had unchanged rosmarinic acid contents as compared to unfrozen controls. The same result was obtained when different cooling rates were used. Cultures cooled at a rate of –10°C/min produced survival rates of 20 per cent compared to the 40 per cent rate achieved under optimum conditions; they still retained their capacity for rosmarinic acid formation. Reuff (1987) also compared the growth and rosmarinic acid production of control cultures and culture after successive freezing and recovery cycles. All recovered cultures showed the same growth and production characteristics as the controls. Stability was also demonstrated for recovery after different storage periods in liquid nitrogen. This work clearly shows that cultures were stable even under sub-optimum freezing conditions, or after successive cryopreservation cycles. Unfortunately no data are available regarding subsequent freezing studies performed under sub-optimum conditions.

Biotin

Similar conclusions regarding the successful application of cryopreservation can be made from freezing experiments with biotin producing callus cultures. Watanabe et al. (1983) published a successful cryopreservation method for biotin-producing callus cultures of *Lavandula vera*. They immersed small pieces of callus taken from the logarithmic growth phase in liquid medium and added an equal volume of a solution of 20 per cent D-glucose and 10 per cent DMSO, which was gradually applied over one hour. Cells were frozen with a rate of –1°C/min to –40°C and then immersed into liquid nitrogen. Although the biotin content of recovered cells was sometimes higher and sometimes lower compared to that of the initial calli, no selection for producing or non-producing cells could be observed. Using the same freezing method Kuriyama et al. (1990) later improved cell recovery considerably in this system by adding activated charcoal to the regrowth medium. These results also indicate that in the early experiments of Watanabe et al. (1993) the cell cultures could be preserved under sub-optimum freezing conditions, without losing their biosynthetic capacity.

Indole Alkaloids

The *Apocynaceae* are a plant family of great pharmaceutical interest and, apart from the cardenolide containing *Oleander* and *Strophanthus*, *Catharanthus* and *Rauvolfia* plants, they produce indole alkaloids which are of importance because of their pharmaceutical uses. Examples include: the heart antiarrhythmic ajmaline and the antihypertensive agent, reserpine. Because of their highly complex chemical structure, indole alkaloids are still not amenable to chemical synthesis.

Nevertheless, the major reason for intensive studies on *Catharanthus* cell cultures is because of the anticancer drugs vincristine and vinblastine. These dimeric alkaloids are only present in very low concentrations in *Catharanthus* plants and they belong to some of the most expensive groups of pharmaceuticals produced. Unfortunately, these compounds could never be isolated from cell cultures; however, although no process of commercial interest could be finally realized, cell lines derived from *Catharanthus roseus* plants belong to some of the most intensively studied plant cell cultures.

Many *Catharanthus* cell cultures contain low levels of monoterpene indole alkaloids and major components include ajmalicine, serpentine and catharanthine. The formation of these compounds can be increased by cell selection. Deus-Neumann and Zenk (1984) were able to increase the levels of ajmalicine and serpentine from 14 mg/l to >350 mg/l using this technique. But the same authors also showed that the high production rates drop if the selection pressure is not continuously maintained. Thus, cryogenic storage is important for the conservation of high producing cell lines of *Catharanthus* and many 'classical studies' regarding the assessment of cryopreservation for maintaining biosynthetic capacities have been performed using this species. Correspondingly, many papers report the successful cyropreservation of *Catharanthus roseus* cell cultures and most have used the standard protocol of Withers and King (1980). Although these reports do not always present data on product formation after cryogenic storage, those that did present findings of extraordinary importance.

In their first attempt to cryopreserve cells of *Catharanthus roseus* selected a non-producing cell line, for the reason that this cell line consisted of very small and dense cells which contained many small vacuoles, but were lacking one large central one. The authors pre-cultured their cells for 24 hours in 5 per cent DMSO and then increased the concentration to 7.5 per cent DMSO for cryoprotection. After one hour of incubation, the cells were frozen using rates of 0.5 to 1°C/min to –40°C and immersed into liquid nitrogen. By this method they achieved a 50 per cent survival rate compared to untreated and unfrozen control cells. They then demonstrated that mitotic index as well as the frequency distribution of the DNA content of re-grown cultures was unchanged by cryogenic storage. Nevertheless, microscopic observations showed that the sub-cellular structure of the cells, re-grown after storage had changed as the small vacuoles had fused to form one large central vacuole.

In a later study, Chen et al. (1984a) found that a modified method (using only one hour preculture) was not successful for the preservation of a high alkaloid producing cell line. However, in contrast to the previous non-producing cell line, this one consisted of much larger cells which had less dense cytoplasm and a huge central vacuole. These authors carried out further investigations to study the mode of action of the cryoprotectants. By NMR techniques they measured the amount (percentage) of water that remains unfrozen in a solution at a given temperature in the state of equilibrium. By using these techniques Chen et al. (1984a) measured the percentage of unfrozen water in the medium and the cryoprotective solutions as well as in mixtures of these solutions which contained cells. They found that the addition of cryoprotective substances increased the level of unfrozen water in the mixtures and established a simple equation between cell survival and the percentage of unfrozen water in a suspension at a given temperature. Cell survival positively correlated with the percentage of unfrozen water. They demonstrated that, using the same cryoprotective solution, the percentage of unfrozen water was lower in the alkaloid producing as compared to the non-producing strain. That this result could not be mimicked by salt solutions showed that cryoprotection is not simply an osmotic effect. It was also shown that the integrity of the membranes had a considerable influence on the percentage of unfrozen water that was moderated by the application of cryoprotective solutions. Nevertheless, by improving their methods, even alkaloid producing strains could finally be manipulated to survive cryopreservation. For the preservation of alkaloid forming cultures, cells were pre-cultured on sorbitol containing agar medium for four days and a mixture of sorbitol (1 M) and DMSO (5 per cent) was applied for pretreatment and cryoprotection. In a subsequent paper Chen et al. (1984b) reported the cryopreservation of three different alkaloid producing cell lines. For the non-producing cell line pretreatment and cryoprotection with DMSO only was sufficient; however, the alkaloid producing strains had to be pre-cultured for 20 hours in medium supplemented with 1 M sorbitol. A mixture of 1 M sorbitol and 5 per cent DMSO was used as cryoprotectant and conditions for programmed freezing and thawing were the same for the producing and non-producing strains. Washing after thawing was performed with the non-producing strain but this eventually turned out to the detrimental for the alkaloid producing strains.

Analysis of the alkaloid content of recovered samples revealed that the total amount of alkaloids formed and the patterns of their product distribution (although data on the alkaloid pattern are only expressed qualitatively as ++ or + for different substances) are not altered by the cryogenic pretreatments or freezing. These results show that although the cryopreservation of non-producing strains is more difficult, freezing of alkaloid-producing strains by improved methods does not specifically select for non-producing cells. Unfortunately, no efforts were made to describe the homogeneity of the investigated cell lines.

Mannonen et al. (1990) used exactly the same method as Chen et al. (1984b) for the cryopreservation of their alkaloid producing cell line of *Catharanthus roseus*. They analyzed the formation of catharanthine and ajmalicine in recovered cells (as they had already done for *Panax ginseng*) and they demonstrated that freezing preserves the biosynthetic capacity of cells far better than continuous subculturing (for 12 months) or medium-term storage under mineral oil (for six months). The ajmalicine content of cells recovered from freezing dropped to 20 per cent of the initial value, and the catharanthine content slightly increased. Unfortunately, Mannonen et al. (1990) did not describe the shape of their cells. From the work of Chen et al. (1984a, 1984b) it cannot be ascertained as to whether it is levels of alkaloid production or simply the shape of cells which makes certain lines more difficult to cryopreserve. Later studies by Suk Weon Kim et al. (1994) demonstrate a clear relationship between alkaloid production and the shape of cells in *Catharanthus* cultures. They isolated several cell lines from the same initial culture and found that alkaloid production is dramatically increased when the cell aspect ratio (cell length : cell width) exceeds a certain threshold. From these results it can be expected that all high producing cells are of the elongated type in the case of *Catharanthus* and therefore these may be more difficult to cryopreserve.

Berberine

Berberine belongs to the protoberberine alkaloids and although several alkaloids of this type are of pharmaceutical importance most attention has been paid to berberine; it is widely used, especially in the Asian market. *Berberis*, *Coptis* and *Thalictrum* cultures are the main systems which have been investigated for protoberberine production. Although protoberberines are formed by cell cultures of several plants and sometimes in high quantities, spontaneously high yielding strains can also be established by cell selection from *Coptis japonica* cell lines.

Cell cultures of *Berberis wilsoniae* have been investigated for berberine production by Reuff (1987) who made great efforts to cryopreserve these cell cultures. She tried almost all approaches for cryopreservation which were known at that time, including different pre-culture methods, cryoprotectants and freezing methods. Although she was able to increase survival rates to 25 per cent, she never achieved re-growth of thawed cultures. Reuff (1987) found that berberine leaking out of damaged cells was not toxic, but other low molecular weight and thermostable compounds from autoclaved cell sap of *Berberis* cultures were toxic to *Berberis* cells. Thus, these compounds produced by damaged cells may kill the few cells surviving cryopreservation after thawing.

Stability of a Biosynthetic Capacity: Cardiac Glycosides

Even in modern medicine, the drugs digoxin and digitoxin still belong to the most widely used and most expensive groups of plant pharmaceuticals. This has stimulated interest to explore *in vitro* techniques as a means of improving the production of the drugs, or even to produce them using cell culture bioreactors. Unfortunately, the corresponding glycosides are not formed in dedifferentiated cell cultures of *Digitalis*. Nevertheless, cell cultures of *Digitalis* are able to perform biotransformations of these compounds and the less important digitoxin can be transformed to the more valuable digoxin. Efforts have been made to use this biotransformation capacity commercially. Another transformation based on the glycosylation of digoxin, digitoxin and gitoxin has also been investigated.

These transformation reactions have also been used by Diettrich et al. (1982) as 'markers' to prove the stability cryopreserved cell cultures of *Digitalis lanata*. For programmed freezing of the cells mannitol was used as the pre-culture osmotic, for one week. A mixture of sucrose, glycerol and DMSO yielded the best cryoprotective results. The freezing rate was between -0.5 to -2°C/min and -60°C and the best survival rate obtained by this method was 51 per cent. The authors investigated the occurrence of membrane damage by microscopy and they also tested the capacity of the cells to glycosylate digitoxin, gitoxin and digoxin, three passages after recovery of the frozen samples. All products (measured only for digitoxin) were transformed and the transformation rate of the cultures recovered after cryopreservation was not changed. The high similarity of the initial and the re-cultivated strains was also demonstrated by measuring the frequency distribution of the DNA content of the cell nuclei by microdensitometry.

A similar freezing method was used by Seitz et al. (1983) for *Digitalis* cell lines. They used a higher mannitol concentration (6 per cent) for a shorter period of time (three days). For cryoprotection a slightly different mixture of glycerol, sucrose and DMSO was used and the freezing rate was -1 °C/min and they achieved a survival rate of over 50 per cent. For testing stability, they measured the transformation of β-methyldigitoxin to β-methyldigoxin and the transformation rate was unchanged in cultures recovered from cryopreservation, compared to unfrozen controls. Small scale and large scale 20 l bioreactors were used for these studies; this was the first report which tested the retention of a biosynthetic capacity of cryopreserved cultures under large scale conditions. It was thus remarkable that stability of *Digitalis* strains after cryopreservation was demonstrated by measuring two different biotransformation reactions, one of them even under large scale conditions. In addition the frequency distribution of DNA content in a cell line was measured and found stable.

Transformed Root Cultures for Secondary Metabolite Production

The fact that valuable compounds (such as atropine) found in plant roots are sometimes not expressed in differentiated cell cultures led to the idea of producing these with root cultures obtained from transformation with *Agrobacterium rhizogenes*. These cultures normally exhibit the same metabolism as normal roots and they can grow without application of plant hormones. Although the problem of generating genetic instability in differentiated cultures is less risky than for dedifferentiated cell lines a preservation method is important to ensure the stabilization of the transgenes. Benson and Hamill (1991) showed that secondary product forming hairy root cultures are also amenable to cryopreservation. They preserved transformed root cultures of *Beta vulgaris* and *Nicotiana rustica*. The simple ultra-rapid freezing of small root tips of 2-4 mm on hypodermic needles was successful, however, for practicability of storage they also used programmed freezing. In contrast to apical meristems, a preculture period of excised root tips turned out to be disadvantageous. Although high survival rates were achieved (46 per cent for *Beta vulgaris* and 83 per cent for *Nicotiana rustica*) the capacity for root conversion was less and a high degree of variation among replicates occurred throughout the whole procedure. Culture age before freezing and the hormone treatment after recovery had a remarkable influence on root regeneration. Nevertheless, for regenerated roots the post-freeze stability of the T-DNA could be demonstrated as well as stability of secondary metabolite formation. For *Nicotiana rustica* roots the ratio of nicotine/anatobine as well as that of nicotine/nornicotine was measured and the betaxanthin and betacyanin levels for *Beta vulgaris* roots showed stability in synthesis patterns after cryogenic storage.

CONCLUSIONS

Although the use of cryopreservation for certain secondary product producing plant cells, tissues and organs is still far from being applied on a routine basis, the first technique that could be addressed as a 'standard method' was the programmable freezing protocol developed by Withers and King (1980). From that time, many different approaches for the cryopreservation of plant cells have been devised. Thus, apart from using controlled rate freezing, today's techniques include: ultra-rapid freezing, vitrification and encapsulation/dehydration. However, most research of dedifferentiated cell cultures which is considered to be of biotechnological importance has been carried out by using programmed, controlled rate freezing. The main reason is that most cryopreservation studies using secondary metabolite producing cell lines were performed during the period 1980 to 1990, at a time when programmed freezing was still the method of choice for plant cell conservation. Other approaches, such as vitrification, were still under development at this time. Another reason for the dominance of the programmed freezing methods is that they still offer the most practical way of handling large numbers of suspension cells.

In almost all cases tested, cryopreservation is able to conserve biochemical stability, and maintain production rates of secondary compounds in cells which have been cryoconserved; similarly biotransformation rates are also stable.

In many cases the stability of other characters such as growth rate, mitotic index, and the frequency distribution of DNA also accompanied positive biochemical stability assessments of the cryopreserved cell lines. The danger of an undesired cell selection after cryogenic storage seems to be less important than expected. It is even more remarkable that on careful analysis of the reports on cryopreservation procedures, in some cases it has been shown that even the application of sub-optimum storage methods did not change the important characters of the preserved cell lines.

However, another important consideration has to be made. In most of the cases where cryopreservation was successful in maintaining biosynthetic capacity, the work was done on cell lines for which the production of the desired compounds normally occurred spontaneously and in high quantities in cell cultures. This is true for ginsenoside production of *Panax ginseng* cells, rosmarinic acid production in *Coleus blumei* cells, diosgenine production in *Dioscorea* cell lines and for the biotransformations of *Digitalis* cells. Since all cultured cells seem to produce these compounds, cell selection would be of no consequence for product formation. In the case of *Catharanthus* cells it is, however, obvious that non-producing and producing cell lines react differently to cryopreservation. It is remarkable that in the case of these cell lines a specific cell shape seems to be linked to alkaloid production. The non-producing cell shape is clearly more amenable to cryopreservation than the shape of alkaloid producing cells. This may be reflected by the molecular structure of the membranes. In the future, it may be important to demonstrate more precisely the role of sub-cellular structures, membranes and cell morphology in the formation of the indole alkaloids.

The key enzyme that combines the iridoglucoside secologanin and the amino acid derived tryptamine to form the basic structure of the indole alkaloid strictosidine, may be localized in the vacuole. Further steps of the biosynthesis seem to take place in the cytoplasm, whereas peroxidases oxidizing ajmalicine to the quaternary serpentine could also be localized in the vacuole. Quaternary alkaloids accumulate in the vacuole either by an ion-trap mechanism or even by active transport. These examples show that alkaloid formation requires highly complex membrane structures. It is easily possible that membranes of alkaloid producing cells are much more complex than those of non-producing cells and that they are also more vulnerable to freezing damage.

Even more complex, is the situation for protoberberines. Although of considerable commercial interest in the 1980s *Berberis wilsoniae* cell lines still cannot be cryopreserved. *Berberis* cultures, even non-selected ones, normally form high yields of protoberberine alkaloids. The quaternary forms of these alkaloids are highly cytotoxic because of their intercalation with DNA. Therefore, complex membrane systems have developed for their biosynthesis. Several of the biosynthetic enzymes including the berberine bridge enzyme (BBE), S-tetrahydroprotoberberine oxidase, S-canadine oxidase and columbamine O-methyltransferase are located in subcellular alkaloid-forming vesicles, which are characterized by their specific gravity. These vesicles also contain the same composition of alkaloids as the central vacuole. Alkaloids seem to be synthesized inside these vesicles and finally accumulate in the central vacuole by the fusion of these vesicles with the tonoplast. How these vesicles are formed and which parts of the biosynthetic pathway take place inside these vesicles is still under discussion. They do seem to be specific for alkaloid biosynthesis and their role may be to prevent cytoplasmic and DNA damage by the toxic alkaloids.

Reuff (1987) demonstrated that berberine leaking out of damaged cells does not poison the surviving cells or cause failure of re-growth of cell lines thawed after freezing. Later, it was shown that *Coptis* and *Thalictrum* cells detoxify exogenously applied berberine and exhibit an uptake system following Michaelis-Menten kinetics. The complex membrane structures for the biosynthesis of these quaternary alkaloids protects the cells as long as they are functioning. But it seems to be possible that even partial membrane damage and loss of compartmentalization during the freezing process may lead to an intoxication of otherwise surviving cells by endogenous quaternary protoberberines produced during recovery growth.

It is a pity that no published data are available as yet regarding the cryopreservation of cell lines which have been used commercially (such as *Lithospermum erythrorhizon* cultures for the production of shikonin or the highly regarded *Taxus brevifolium* cultures which will probably be used in the future for taxol production). Finally, it may be concluded that when routine cryopreservation procedures are developed and applied on larger scales in the secondary products sector, it can be achieved with some confidence. The present data indicate that the biosynthetic potential of dedifferentiated cell lines can be best maintained by cryopreservation. This also take into account the expected problems associated with the inherent heterogeneity of many secondary product forming cell lines.

19

GERMPLASM CONSERVATION

The development of useful *ex situ* germplasm collections, whether for genetic resource conservation or in support of crop improvement programmes, requires that plants be collected from various sites around the world and introduced to new locations. A goal of many germplasm repositories is to represent the global genetic diversity of plant genera in their charge. One of the risks associated with collection of plant germplasm, especially from wild sources, is the inadvertent introduction of diseases or other pests along with plants or seeds. Some of the world's most destructive plant diseases have been the result of the accidental introduction of exotic pathogens during the importation of plant materials. Examples of such importations include downy mildew of corn and sorghum caused by *Peronosclerospora sorghi*, white pine blister rust (*Cronartium ribicola*) chestnut blight (*Cryphonectria parasitica*, and more recently karnal bunt of wheat (*Neovossia indica*), sharka disease of *Prunus* sp. and tomato spotted wilt virus. Virus and virus-like diseases pose a special challenge as they often symptomless in infected plants and require specialized tests to determine their presence. As plant breeders attempt to broaden the genetic base of our agricultural and horticultural crops there are increasing efforts to introduce new genetic material from areas of origin of the species involved. These introductions are often land races or wild species that are not closely examined for the presence of germplasm-borne pathogens.

SAFE MOVEMENT OF GERMPLASM

Quarantines have become the primary strategy for preventing the international movement of viruses and other pests along with plant material. Finding a balance between excluding pests through the use of quarantine measures and introducing germplasm for the improvement of agricultural production is often a challenge. The destruction of plant material found to be infected with a pathogen or the many years of testing often required before material can be released from quarantine, has frustrated farmers, horticulturists and breeders anxious to incorporate new genetic resources into their programmes. *In vitro* culture and molecular diagnostic techniques are helping to expedite the detection and elimination of pathogens from plant germplasm. The protection of virus tested clones in certification programmes helps to safeguard valuable plant material, and prevent the reintroduction of pathogens into clean stock. With increased international movement of germplasm the challenge to present introduction of exotic pathogens is also increased. The ease of air travel provides a multitude of opportunities for the introduction of a 'favourite' fruit, vegetable or flower; however, this introduction can lead to serious consequences if an exotic plant pathogen is inadvertently introduced at the same time. It is believed that this is how necrotic strain of potato virus Y (PVY^N) was introduced into eastern Canada, resulting in millions of dollars in losses to the seed potato industry in that region.

Quarantine

Quarantines are the first line of defence against the movement of economically important plant pests between and within countries. Most countries have enacted quarantines to prevent the economic losses that result from the introduction of exotic insects or diseases Introduced pests have been responsible for devastating losses of native plants and cultivated crops. Quarantines are not only used between countries, but are also an important administrative tool for preventing the spread of diseases within countries. In the United States, for example, certain states with pine based timber industries restrict the importation of *Ribes* species, which are an alternate host for *Cronartium ribicola* (white pine blister rusts). The state of Oregon restricts the importation of *Corylus* species (hazelnut) from the eastern two-thirds of North America as well as from several Oregon countries to prevent the further spread of *Anisogramma anomala* Peck. Provinces in western Canada controlled the movement of potatoes from the Maritime Provinces during the early 1990s to protect their seed potato industries from the introduction of PVY^N.

International Cooperation

The International Plant Protection Convention (IPPC) was adopted by a conference of the Food and Agriculture Organization (FAO) of the United Nations in 1952 with the purpose of 'securing common and effective action to prevent the spread and introduction of pests of plants and plant products and to promote measures for their control' (FAO, 1998). As of 1996, 105 countries had agreed to abide by the IPPC. The IPPC directs member countries to develop procedures for issuing phytosanitary certificates, and encourages international cooperation to prevent the spread of pests while minimizing interference with international trade. While most individual countries maintain their own plant quarantine regulations, several regional plant protection organizations have been formed by IPPC member countries in an attempt to provide regional consistency in regulations, and cooperation in the development of plant certification schemes. Contact information for these regional organizations can be found at the FAO website. Three additional regional organizations administer plant protection programmes not associated with the IPPC. The FAO has defined two categories of quarantinable pests. An 'A-1' pest is not yet present in an area and is the most significant from a quarantine standpoint. An 'A-2' pest may be present in an area but is not widely distributed. Quarantine regulations are generally more rigorous for A-1 than for A-2 pests.

Europe

The European Plant Protection Organization (EPPO) makes recommendations to its 39 member countries in Europe and the Mediterranean basin regarding plant quarantine issues related to the movement of commodities.

The 15 nations of the European Union (EU) abide by EU phytosanitary laws, many of which are consistent with EPPO recommendations. EPPO has published a number of certification schemes, for example for strawberries and for fruit trees, with specific suggestions for selection, production and maintenance of nuclear stock, and guidelines for pathogen testing and sanitation.

United States

In the United States, importation of nursery stock, plants, roots, bulbs, seeds and other plant products is controlled by foreign quarantine regulations. The Animal and Plant Health Inspection Service (APHIS), an agency of the US Department of Agriculture, is charged with enforcing plant quarantine regulations. For most plant genera, seeds have fewer restrictions than vegetative materials. Plant material imported for propagation will fall under one of the following categories, depending on the plant species and the country of origin:

Restricted

These plant propagules can be imported by any individual in any quantity but are subject to inspections for quarantine pests.

(a) Many plant species are not mentioned in the US quarantine regulations. Such plants or seeds may be inspected for visible evidence of diseases or insects at the port of entry, and if the propagules appear healthy, they are released to the importing individual.

(b) Individuals must obtain a permit to import other plants. These plants or seeds must be inspected at the port of entry and if there is no sign of insects or disease, they are released to the permit holder.

(c) Plant materials which are at risk of harbouring more hazardous pests must be imported with a permit, inspected at an inspection station, and then grown according to certain 'post-entry' conditions. Post-entry restrictions generally require that plants be grown at an approved site, kept a certain distance away from other plants of the same genus, and inspected by an agricultural official several times during two growing seasons.

Prohibited

Certain vegetatively propagated plants, including many of the fruit and nut crops, cannot be imported directly by private individuals or organizations in commercial quantities. These plants can only be introduced in small quantities through a government approved quarantine facility, where they must be tested for economically important pathogens (especially viruses) before being released for general propagation and dissemination.

International Guidelines

The International Plant Genetic Resources Institute (IPGRI) is another organization associated with FAO which promotes the conservation and use of genetic resources. In recognition of the phytosanitary hazards involved in the international movement of plant germplasm, IPGRI has funded a number of research programmes and conferences to develop appropriate technologies for reducing the risks of moving pathogens with plants. A series of crop-specific handbooks have been published by IPGRI to provide technical guidelines for the safe movement of a number of economically important crops plants.

These guides have especially targeted vegetatively propagated crops which carry a higher risk of carrying virus diseases and they suggest appropriate methods for excluding these diseases during germplasm exchange. Some general recommendations to promote the safe movement of germplasm include:

1. Germplasm should be obtained from a safe (pathogen tested) source.
2. Movement of seed or pollen poses less risk than does movement of plants.
3. *In vitro* cultures pose less risk than does vegetative plant materials potted in soil.
4. *In vitro* material should be derived from pathogen tested sources or tested for viruses known to occur in the country of origin.
5. Transfer of germplasm should be planned in consultation with quarantine authorities and a relevant indexing laboratory.
6. Vegetative material should be subjected to full quarantine measures.

Reviews of successful certification programmes have recently been published for potatoes, grapes, ornamental plants, deciduous fruit trees, citrus, and strawberries. Common features of certified stock schemes involve:

1. Listing the pathogens of concern.
2. Identifying appropriate methods for detecting the pathogens.
3. Implementing strategies for eliminating pathogens from infected plants.
4. Protecting foundation or nuclear stock from reinfection.
5. Verifying that plants are correctly labelled or are 'true-to-type'.
6. Preventing reinfection during subsequent propagation and distribution to commercial producers.

IPGRI recommends the development and use of 'broad-spectrum' virus detection techniques for quarantine indexing. Such tests may not identify specific pathogens, but rather will detect members of a larger group such as all plant phytoplasmas, or all members of a virus family.

Virus Detection

Accurate disease diagnosis combined with sensitive, rapid and early detection of plant viruses is critical for effective management of most crop systems. However, in plant germplasm collections detection of a wide range of viruses, some of which may not be known, is required generally for a small number of samples. There have been several major improvements in virus detection over the past two decades. Serological detection was greatly improved with the introduction of ELISA to plant virus detection in the mid-1970s. This was enhanced further with the development of monoclonal antibody technology and its application to a large number of plant viruses in the 1980s. Similarly, nucleic acid hybridization has been used successfully for detecting many plant viruses and is the preferred method for detection of viroids. Cloning of plant viral nucleic acids and the development of non-radioactive detection methods have increased the polymerase chain reaction (PCR) has greatly improved the sensitivity and utility of hybridization and other nucleic acid based assays. Immunocapture PCR combines the advantages of serology and PCR into a very sensitive method of detection.

In germplasm repositories it is often necessary to test for viruses in accessions that are from diverse geographical locations and that may be infected with unknown viruses. Tests for common viruses known to occur in a genus should be carried out. However, repositories have the added need to determine whether these materials may be infected with yet undescribed viruses. This necessitates the use of broad spectrum techniques such as mechanical transmissions to herbaceous indicators, electron microscopy, double-stranded RNA analysis, grafting onto indicator hosts or looking for viral inclusions.

Monoclonal antibodies, nucleic acid hybridization and PCR provide the potential for the development of diagnostic reagents with desired specificities. Diagnostic reagents that detect all or most members of a virus group would be very useful in germplasm repositories. There are monoclonal antibodies that recognize most members of the potyviridae or multiple luteoviruses and oligonucleotides that can be used in PCR to amplify sequences from most luteoviruses or geminiviruses.

Requirements for Detection and Diagnosis

The requirements for specificity and sensitivity of detection will vary in different situations. Germplasm repositories and clean plant programmes want to ensure that their 'nuclear' material is free of known viruses. In this situation, where much depends on the virus status of relatively few individual plants, several tests should be employed and the virus status of every plant may determined. While serological or nucleic acid-based tests often can be employed for detection of well characterized viruses known to occur in the crop, tests that detect a wide range of viruses are especially desirable. Mechanical transmission to selected herbaceous indicator plants, electron microscopic examination of leaf dips, double-stranded RNA (dsRNA) analysis and/or grafting may be desirable to ensure that the germplasm or 'nuclear' material is virus-free. 'Nuclear' material refers to the few plants that are the basis of all plant material in a clean plant propagation scheme and may also be referred to as mother block, foundation, or elite material. New material being added to virus-free collections may be put through a virus eradication programme prior to testing. This enhances the likelihood that the material will be free of viruses not reported previously in the crop and free of new strains of a virus which may not be detected with available antibodies or nucleic acid probes.

Germplasm repositories often have permits to bring in material to meet their mandate without having the plants go through a plant quarantine station. Therefore, their virus detection programmes should be similar to those of plant quarantine. In plant quarantine it is necessary to ensure that plant material is free of restricted viruses prior to its release to industry or breeding programmes. Quarantine programmes often work with only a few plants of a given genotype but the level of indexing on these few plants is as complete as possible. Maximum sensitivity is the goal of virus detection in quarantine programmes. These programmes often use an agreed upon test for each specific virus. The test may be grafting or mechanical transmission onto specific indicator plants, an ELISA, hybridization assay or PCR for specific viruses. Detection of viruses of woody plants will often require graft transmission tests to indicator plants. When grafting is required for a single virus in a crop it might be more efficient to employ graft transmission tests for all viruses of quarantine significance since the work is already being done for the one virus. Thus, quarantine facilities may not opt for the newer laboratory techniques until they are available for all quarantinable viruses capable of infecting that crop.

Quarantine officials may also require extreme specificity in cases where a severe or resistance breaking strain of a virus occurs in some countries but not others. When there are several sources of plant material it might be easier to destroy all material that is infected with a particular virus rather than trying to determine which strain of a virus is present. There is a resistance breaking strain of raspberry bushy dwarf virus (RBDV) that occurs in Europe and is not known to occur in North America. The only way to differentiate the two strains is to graft onto resistant varieties of raspberry. Both strains are readily detected, but cannot be differentiated serologically and in practice any material being imported to North America from Europe that tests positive for RBDV is destroyed rather than risk introducing the resistance breaking strain of the virus. Recently, reverse transcription combined with PCR (RT-PCR) has been developed to differentiate these two strains of RBDV and as this test is improved it may be possible to reliably detect resistance breaking strains of RBDV.

Serological Detection

Enzyme-linked immunosorbent assay (ELISA) and dot immunobinding assay (DIBA) are currently the most widely used methods of serological detection for plant viruses. There have been relatively few changes in ELISA or DIBA for virus detection since the application of monoclonal antibodies (McAbs). ELISA has been reviewed elsewhere and the general outline will not be covered here. Rather, some of the principles of the assays will be discussed. These assays are carried out on a solid phase (usually plastic multi-well plates, nitrocellulose membranes or filter paper where each component of the test is applied successively and the reaction between virus and antibody is detected by enzymatic hydrolysis of a substrate that results in a colour change or light emission.

Assays carried out on nitrocellulose or filter paper are referred to as DIBA. The DIBA is about as sensitive as ELISA and offers the advantage that sample preparation can be very simple. A few microlitres of sap can be spotted onto the membrane or a freshly cut edge of a leaf, petiole or stem can be pressed gently against the absorbent membrane. The latter approach is referred to as tissue blotting and also provides some information on the location of the virus in the tissue, e.g. phloem. The other advantage of DIBA is that it is readily adapted to field situations and applications in areas with minimal laboratory facilities. The membranes can be taken to the field and plant tissue or insects blotted directly onto the membrane. Both of these assays can be performed without any specialized equipment.

The ELISA procedure is quite adaptable and many variations on the original method have been described. A standardized test protocol should be used in quarantine and certification schemes to ensure consistent results between laboratories or from year-to-year in the same facility. Standardization of an ELISA protocol may include factors such as the manufacturer of a microtitre plate used in the assay, identification of a specific monoclonal or polyclonal antiserum, part of plant to be sampled and at what stage of plant development sampling should take place, list of buffers to be used at each step of the procedure, the length and temperature of each incubation, how long the substrate should develop before absorbance values are taken, and what should be used as positive and negative controls.

ELISA in plant virology usually has two or three steps that use an antibody. The antibody applied directly to the microtitre plate is usually referred to as the trapping or coating antibody, since its purpose is to selectively bind the antigen of interest to the plate. The coating antibody is applied as purified immunoglobulin G (IgG) diluted in an appropriate buffer (usually carbonate or phosphate buffered saline). The second antibody (often the same source of IgG as the coating antibody) is conjugated with an enzyme in the standard double antibody sandwich (DAS) ELISA and is referred to as the conjugate or detecting antibody. In triple antibody sandwich (TAS) ELISA, the coating is done in the same manner as in DAS-ELISA but the second antibody (primary or detecting antibody) is specific for the antigen of interest and produced in a different animal than the trapping or coating antibody. The primary antibody is then followed by a conjugated antibody (secondary antibody) that is specific for antibodies produced by the animal that was the source of the primary antibody. For example, if the primary antibody was a McAb produced in a mouse, then the secondary antibody might be a rabbit-anti-mouse conjugate. We prefer a conjugate that is made in the same host species as the coating antibody to prevent cross reaction between conjugate and coating antibody.

When using monoclonal antibodies (McAbs) for virus detection or diagnosis, the substrate can usually be incubated overnight to increase the sensitivity of the assay. In this way, McAbs can be used to detect virus in individual aphids using a standard ELISA protocol. In our laboratory, we routinely take absorbance readings 1-2 hours after adding substrate and again after overnight incubation at room temperature. The most important aspect of developing an assay is to maximize the absorbance of infected/healthy samples. A good polyclonal or monoclonal antiserum can be used to develop an assay that does not require statistical analysis to differentiate between known positive and negative samples. If such a test is available, samples that give borderline results should be retested. It must be remembered however, that even with the best of detection methods, recently infected samples may well give a borderline or negative result. The application of statistics to determine when an absorbance value in ELISA represents a positive result has been present elsewhere.

Nucleic Acid Based Assays

Detection of pathogens by nucleic acid hybridization is based on the specific pairing between the target nucleic acid sequence (denatured DNA or RNA) and a complementary nucleic acid probe to form double-stranded nucleic acids. Thus, either RNA or DNA sequences may be used as probes. For detection of plant viruses, hybridizations are usually carried out on solid filter supports where the target nucleic acids are immobilized and the labelled nucleic acid probe is allowed to hybridize to them. Nitrocellulose and charged nylon are the most commonly used filters for hybridization. Nylon based filters are easier to handle and can be rehybridized several times. The use of nucleic acid hybridization assays for virus detection has been reviewed recently. The greatest improvements in hybridization assays in recent years have been the advances in non-radioactive detection systems. Several methods of labelling nucleic acids are available. Incorporation of modified nucleotides such as biotin-11-UTP, or digoxigenin tagged UTP, can then be detected by streptavidin or an anti-digoxigenin antibody, respectively. Digoxigenin-labelled dUTP appears to be the most widely used non-radioactive tag for labelling probes for detection of plant viruses. Digoxigenin is linked via a spacer arm to dUTP and then incorporated into probe with the same enzymes used to make ^{32}P-labelled probes. Antibodies specific for the digoxigenin and conjugated to an enzyme (usually alkaline phosphatase or horseradish peroxidase) are then used to complex with the bound probe. Either a precipitating or a light emitting substrate is then used to visualize the presence of the probe. Many biotechnology supply companies now marked kits for tagging probes with non-radioactive labels.

The sensitivity of detection of plant viruses by nucleic acid hybridization is roughly similar to that of ELISA. Hybridization assays were more sensitive than ELISA with subterranean clover stunt virus (SCSV) when purified virus was used; however, when infected tissues were used, ELISA was found to be more sensitive: the pathogen was detectable at a sap dilution of 1/625 while hybridization assays had a dilution limit of 1/125. The sensitivity of nucleic acid hybridization and ELISA were similar with barley yellow dwarf virus (BYDV) and cherry leaf roll viruses.

Another consideration in the choice of detection procedure, is the ease of carrying out an assay. If two methods of detection are sensitive enough to meet the needs of an experiment, then the method that is simpler to carry out will probably be used. The expertise of the worker will determine which test is simpler. Someone who has extensive experience working with nucleic acids will be more inclined to use PCR or hybridization assays rather than ELISA; the converse is true for someone who is more familiar with serology than nucleic acid based tests.

There are many viruses of woody plants that have only been described in terms of symptomatology. It is likely that nucleic acid based detection will be available before serological methods are developed for these viruses. Developments in cloning viral specific dsRNAs of plant viruses make it possible to readily make cDNA from dsRNA templates. Thus, for viruses where dsRNA can be extracted from diseased tissues, probes for nucleic acid hybridization or sequence information required to develop a PCR test are realistic short-term goals. Once the sequences are known, it will be possible to prepare antibodies to the coat proteins of these viruses for use in serological assays.

In the case of phytoplasmas, several groups have described universal oligonucleotides for amplification of phytoplasma specific DNA. Since this group of pathogens cannot be cultured, are not easily transmitted by grafting or mechanically, and can be difficult to detect by microscopy, the PCR based test is the preferred method for detecting phytoplasmas.

Detection Based on more Traditional Methods

Serological or nucleic acid based diagnostic tools are not available to detect many of the viruses of woody plants because it has not been possible to purify the viruses. It is still necessary to carry out graft transmissions or mechanical inoculations onto herbaceous hosts for these viruses. Attempts to improve quarantine and certification schemes for tree fruit or small fruit crops is limited because these tests are labour intensive, and especially with grafting, require substantial amounts of space to grow test plants. It may be several years after grafting before the results of the test are obtained.

The introduction of new germplasm from collections made in the wild presents the possibility that undescribed viruses will be encountered. New viruses encountered will not be on any quarantine list but may still pose a biological risk. This material should be assayed with a broad spectrum test such as mechanical transmission, grafting or dsRNA analysis. Quite often we tend to be concerned about the viruses with which we are familiar and ignore those that have not yet been described.

Significance of a Test Result

In situations where a test result has significant biological or economic impact, one should use more than one type of test to confirm the presence of absence of a virus. A recent incidence of PVY^N in Canada and its implications for shipping seed potatoes to the USA is an excellent example of where test results had considerable economic impact and a confirmatory test could potentially have saved a lot of money and possibly prevented lawsuits. In this case the effect of mixed virus infections confounded the bioassay results on tobacco leading to a large percentage of false positives. As the movement of agricultural products between countries increases, quarantine restrictions based on plant pathogens will continue to be important. For example, several *Prunus* spp. were found to be infected with virus isolates which cross-reacted with plum pox potyvirus (PPV) antisera in ELISA. In this instance, the impact of the test result was very significant since plum pox is an A-1 quarantine status virus not known to occur in North America. Plum pox is a member of the potyviridae that produce diagnostic inclusions in infected hosts. The prunus virus isolates did not induce these inclusions; their coat proteins were a typical of members of the potyviridae and RT-PCR tests with oligonucleotides specific to the 3' or 5' non-coding regions of plum pox virus did not amplify any fragments. It is now thought that these prunus virus isolates are not plum pox virus, nor are they members of the potyviridae despite their reaction with antisera specific to plum pox virus. Reliance solely on the initial test results would have had very serious implications.

Another important consideration when using diagnostic tests that are not based on biological activity is the significance of the result. For example, many people are aware that PCR has been used to amplify DNA from insects embedded in amber for millions of years. This positive 'test' for insect DNA does not show that the insect is living. A similar situation could arrive when using laboratory tests to index plants for viruses. It has been shown that biologically active prunus necrotic ringspot virus is slowly lost from *Prunus pennsylvanica* seed. Would ELISA, PCR or hybridization give positive results when biologically the virus was no longer seed transmitted? In the case of biological tests, it is also important to know that the symptoms observed are due to the virus in question rather than a mixed infection as was the case with some of the PVY^N testing mentioned above.

The legal implications of a decision to refuse a shipment of plant product based on a positive test that does not consider biological activity is an issue that needs to be considered. If a shipment of grain that has been fumigated with methyl bromide is assayed for a specific fungal pathogen by PCR, it would be possible to get a positive result based on non-living fungal tissue. Similarly, the results of a positive ELISA test for plum pox potyvirus could be used to turn back a shipment of

Prunus planting stock, when in fast the material was free of plum pox potyvirus. We must remember to consider the biology of the pathogen and host rather than rely on a band in a gel or yellow colour development in a well of a microtitre plate.

Production of Pathogen-free Plants

Most plants that are vegetatively propagated from a virus-infected plant will be infected with the same viruses. These clonal plant populations can become infected with additional viruses if exposed to virus-carrying vectors such as nematodes or aphids. Many heirloom fruit cultivars have been clonally propagated for centuries and are universally virus infected, although the viruses may be latent or symptomless during much of the growing season. Pathogen-free plants can sometimes be identified by careful indexing, but often the only way to obtain healthy foundation stock for important clonally propagated plant cultivars is to subject them to virus-elimination therapy.

Exposure of growing plants or plant parts to elevated temperatures (heat therapy or thermotherapy), to apical meristem culture, or to antiviral chemicals (chemotherapy) has been used to eliminate viruses from infected plants. Improved virus elimination can often be achieved by combining these various techniques. The treated plants generally are not cured during therapy, but rather new plants are propagated from shoot tips or apical meristems following treatment. Plant tissue culture offers a convenient system for exposing plants to controlled temperatures or to controlled concentrations of antiviral chemicals. *In vitro* therapy also provides already-sterile plant tissue from which meristems can be dissected with no need for additional surface sterilization.

Heat Therapy

While the exact effect of heat therapy on plant viruses is not well understood, it is known that replication of many viruses is significantly reduced at elevated temperatures. The production of virus-encoded movement proteins and coat proteins may also be temperature sensitive. These proteins are involved in cell-to-cell movement of viruses through plasmodesmata and long distance movement through the plant vascular system. Disruption in the production or activity of these proteins may also play a role in the effectiveness of heat therapy.

Nyland and Goheen (1969) reviewed the early history of using heat to eliminate viruses and other pathogens from infected plant material. Brief exposure of plants or plant parts to hot water at temperatures ranging from 30° to 70°C has been used successfully to eliminate many pathogens from infected plants, but nearly all of the pathogens eliminated by hot water baths were later discovered to be phytoplasmas and not viruses. Hot water may be a useful technique for sanitizing plant parts to eliminate arthropods, fungi, bacteria and phytoplasmas, but it has not been particularly useful for eliminating viruses.

Most modern virus therapy programmes involve growing whole plants or *in vitro* cultures at temperatures close to the threshold of normal plant growth. For most plants this is between 38° and 40°C. Mink et al. (1998) reviewed the effect of elevated temperatures on viruses and on plant physiology. Reduced synthesis of RNA at 40°C has been shown for several viruses, including tobacco mosaic virus (TMV) and cowpea chlorotic mottle virus (CCMV). At elevated temperatures synthesis of ssRNA stopped immediately for both of these viruses and synthesis of dsRNA stopped more gradually. When plants were returned to normal temperatures of 25°C, resumption of viral RNA synthesis lagged behind plant RNA synthesis by 4-8 hours for CCMV and by 16-20 hours for TMV.

Infected plants may be subjected to constant therapy temperatures, however alternating temperatures are commonly used to improve host survival. The lag in resumption of viral RNA synthesis behind that of the host plant may explain why alternating-temperature heat therapy is perhaps more successful than constant-temperature therapy. Plant survival is improved during alternating-temperature therapy. We have successfully eliminated many viruses from numerous fruit, nut, and oil crop genera (*Corylus*, *Fragaria*, *Humulus*, *Mentha*, *Mespilus*, *Pyrus*, *Rubus*, *Ribes*, *Sorbus*, *Vaccinium*) by growing either potted plants or *in vitro* plantlets at temperatures that alternate every four hours between 30° and 38°C. After two to four weeks of heat therapy, apical meristems are dissected and established *in vitro*. Once these meristem derived plants are rooted and established in soil, they are allowed to go through a natural dormant period and retested for viruses during the following growing season. The dormant period allows viruses, that may have been reduced to undetectable levels by therapy, to build up in the plant prior to indexing.

While typical virus therapy involves growing an infected plant at elevated temperatures prior to meristem culture, cucumber mosaic and alfalfa mosaic viruses have been eliminated by subjecting dissected meristems to heat therapy after they were removed from infected plants.

Cold therapy rather than heat therapy, followed by apical meristem culture, has successfully eliminated several viruses from infected plants and cold therapy has been particularly effective in the elimination of viroids, some of which are quite resistant to elevated temperatures. Plants or *in vitro* cultures are typically grown at temperatures between 4° and 7°C for one to six months prior to removal of meristems to eliminate viroids including potato spindle tuber, chrysanthemum stunt, and apple scar skin viroids.

While there seems to be little agreement on the importance of humidity or light levels during therapy, it has been shown that elevated CO_2 enhances survival of some plant species at elevated temperatures. Blueberry (*Vaccinium* spp.) plants, which normally are difficult to keep alive at 38°C were able to survive for extended periods at 40°C when the CO_2 level was increased to 1200 ppm, 3-4 times the normal concentration. We have observed improved survival of the infected host during *in vitro* heat therapy if the plantlet is cultured in a heat-sealed gas-permeable plastic pouch rather than a test tube or other culture container. These plastic containers allow some gas exchange but are impervious to the exchange of moisture. Improved plant condition may be due to a decrease in fungal or bacterial contaminants, better water retention in the medium, or elevated CO_2 levels in these containers.

Viruses vary in their susceptibility to heat therapy. Apple mosaic ilarvirus, blueberry scorch carlavirus, and several of the mosaic viruses of raspberry are readily eliminated following therapy times of only 10-20 days and in the case of apple mosaic, following propagation of relatively large shoot tips over a centimetre in length. Other viruses such as raspberry bushy idaeovirus, tobacco streak ilarvirus and apple stem grooving capillovirus persist in some meristems smaller than 0.5 mm dissected after 3-4 weeks of heat therapy.

Meristem Tip Culture

Virus elimination has been documented for a number of virus-host combinations following prolonged *in vitro* culture of the infected host plant, and in many additional instances following excision and culture of apical meristems. Faccioli and Marani (1998) recently reviewed the various roles that *in vitro* culture has played in elimination of viruses from infected plant material. The mechanisms for virus elimination during routine tissue culture are uncertain, however the presence of plant growth hormones, the physiological response of explants to repeated injuries, the induction of inhibitors or the disruption of enzymes needed for virus replication are several possibilities. Cherry leaf roll virus and arabis mosaic virus for example, have disappeared from infected plants after several months of growth *in vitro*, yet tobacco ringspot, cucumber mosaic and potato virus X are persistent.

It has been suggested that many viruses are unable to infect the apical meristem of a growing plant and that a virus-free plant can be produced if a small enough piece of apical tissue is propagated. Facciolo and Marani (1998), however, cite a number of studies where electron microscopy and fluorescence-linked antibodies have documented the presence of more than a dozen different viruses in apical dome tissue of assorted plant species. Yet plants regenerated from these infected meristems are often free of the viruses. The larger the size of an excised meristem the better the chance that it will survive *in vitro* culture, and the smaller the meristem size, the more likely it will be virus-free. The goal of many virus elimination programmes is to grow a meristem consisting of the apical dome and a pair of leaf primordia. Depending on the plant genotype, this shoot tip should be between 0.2 and 0.8 mm in length. Distribution of a virus within a plant may be uneven, especially towards the shoot tips, and a virus-free plant may be produced by random propagation of enough buds or shoot tips. Experience with a particular crop and the viruses that infect it will determine the size of the meristems that can be established *in vitro*, and how many meristems must be grown to have a reasonable probability that one of them will be virus-free. A single, virus-free, true-to-type plant is all that is necessary to produce a population of healthy plants.

Some woody plants are difficult to establish in *in vitro* culture from meristems, or difficult to root. These difficulties have been overcome in *Citrus*, *Malus*, and *Prunus* by 'micrografting' shoot tips or apical meristems onto *in vitro* grown seedling rootstocks, which are later transplanted to soil after the grafts have become established. Pears which grew easily from *in vitro* meristems, but which were difficult to root, could be removed from tissue culture when they had elongated to about 1 cm and successfully cleft-grafted onto potted seedlings. Selection of rootstock seedlings with distinct leaf colour or morphology aids in differentiating micrografts from rootstock sprouts as they grow out.

Although it is possible to eliminate viruses from plants following meristem tip culture alone, this procedure is almost always combined with heat therapy or chemotherapy to increase the likelihood of success.

The combination of heat therapy followed by apical meristem culture has become the foundation for many virus elimination programmes at germplasm repositories, research institutes and commercial plant nurseries around the world.

Chemotherapy

Unlike fungi and bacteria, viruses cannot be eliminated from infected plants by protective chemical sprays. Several chemicals, however, can be used in the same manner as heat to inhibit the replication or movement of viruses thus producing a region of virus-free tissue. The chemicals can either be sprayed on growing plants or incorporated into tissue culture media. Following a period of chemotherapy, shoot-tips or apical meristems are excised and propagated as in heat therapy. While it may be possible in some instances to cure an infected plant, the cost of these chemicals precludes the application of chemotherapy to field trees.

Many of the antiviral chemicals that have been used for plant virus chemotherapy are synthetic nucleotide analogues that had previously shown some effectiveness against animal viruses. The concentrations that are required during chemotherapy to inhibit virus multiplication are very close to the concentrations which are toxic to the host plant. The guanosine analogue ribavirin (Virazole 1-D-ribofuranosyl-1,2,4-triazole-3-carboxamide), and the uracil analogue DHT (5-dihydroazauracil) are two substances which are particularly effective at inhibiting many different plant viruses. Hansen (1988) reviewed applications of chemotherapy to virus infected plants. In early studies, antiviral substances were injected into stems or applied to whole plants as foliar sprays or root drenches. More recent work involved the incorporation of antiviral materials into tissue culture media leading to the elimination of viruses from many plants including peanut (*Arachis*), orchid (*Cymbidium*), potato (*Solanum*), and strawberry (*Fragaria*). Tissue culture methods permit more control over the concentration of chemical agents, the period of exposure, and the ability to combine chemotherapy with heat therapy or apical meristem culture. Combinations of more than one chemical are also more easily evaluated using tissue cultures.

The modes of action of the various plant virus chemotherapies are not well understood. Ribavirin, perhaps the best studied, has been shown in animal viruses to interfere with capping at the 5' end of viral mRNA and many of the ribavirin-sensitive plant viruses also have a 5' capped mRNA. While chemotherapy is effective for eliminating a diverse array of viruses from many different plant species, the chemicals used, the effective concentrations, and the phytotoxicity levels vary greatly depending on the plant genotype. The possibility of genetic mutations when plants are exposed to antiviral chemicals poses risks that have not been adequately studied. Heat therapy poses a much lower risk of genetic change to treated plants, and the procedure is consistent regardless of the virus or the plant host species. Chemotherapy may supplement the other methods of plant virus elimination, but is not likely to replace them, except in situations where viruses are not easily eliminated by heat and meristem culture.

Not all plants resulting from therapy will be pathogen-free. Each plant must be subjected to follow-up indexing to confirm the virus status, and plants should be grown out and examined for trueness to type before being offered for commercial production. There are abundant opportunities for plants to become mislabelled during the various stages of propagation, heat therapy, *in vitro* culture, and re-establishment. Concerns have also been raised about the possibility of genetic mutations or selection of somaclonal variants especially following *in vitro* culture procedures which involve certain plant growth hormones. Such changes are less likely if regeneration from undifferentiated tissues or callus is avoided during *in vitro* culture. Heat therapy and apical meristem culture not only can eliminate viruses that have been previously detected, but may also eliminate exotic or latent viruses whose presence is unknown, as well as contaminating organisms such as bacteria, fungi, or phytoplasmas.

CONCLUSIONS

Germplasm collection present a unique set of problems with regard to prevention of movement of plant viruses. Due to the nature of the plant material collected, there is always the possibility of introducing completely unknown viruses. A more complete set of virus tests needs be carried out on these plants compared to other virus testing programmes such as certification programmes dealing with a known set of viruses. Since germplasm collections deal with a broad range of genetic materials, sensitivity to thermotherapy or chemotherapy, and the ability to culture the plants *in vitro* may be more variable than in certification programmes where the genetics of a crop is more uniform.

Once pathogen-free germplasm is identified, whether through indexing of collected accessions, or through a combination of therapy and indexing, the plant material should be protected from reinfection. In conventional plant gene banks and certification programmes this involves isolating growing plants from virus vectors. Potted plants maintained in insect-proof enclosures will be protected from spread of viruses by insects and nematodes. Removal of flowers and exclusion of pollinators such as honeybees will prevent the movement of pollen-borne viruses. Conservation of clonal plant germplasm *in vitro* should assure that plants are protected from becoming infected with pathogens, including viruses, provided the source of the *in vitro* plants is virus-free.

20

Prospects and Future Challenges

The rapid development of techniques for plant cell and tissue culture over the last three decades has had a major impact on many areas of plant science. During this period, the number of technical reports detailing protocols for the successful application of tissue culture technology to a wide range of species has dramatically increased. It is now possible to regenerate plants from a range of tissue explants from many plant species, including most of the important crop, fruit and ornamental species. More recently, developments in the science of plant cell and tissue culture have combined with rapid advances in molecular biology to drive a new revolution in plant biology and to launch the field of plant biotechnology. Some remarkable breakthroughs have been achieved and many procedures have been devised that are currently valuable tools for basic research or have been successfully exploited in commercial applications.

The Future

It can be anticipated that plant biotechnology will continue to evolve in response to the immense challenge that is presented by the vast range of biological diversity in plant species. The advancement of established areas and the exploration of new directions will require the constant development and refinement of cell culture technology. New possibilities making use of the information obtained from getlonle sequencing of crops like rice, the application of technologies such as genomics, proteomics and metabolomics are being explored, and there exists the likelihood of impressive breakthroughs in the foreseeable future.

Improved Regeneration

The combined techniques of tissue culture and molecular biology have advanced to the point that transgenic plants can be created by the introduction of genes from almost any source. Furthermore, the exploitation of these technologies has led to the commercialization of transgenic crops. However, as a general strategy, efficient transformation and regeneration remains a feat of skill because each species or genotype requires unique culture conditions. The manipulation of species that have not been previously transformed or cultured requires either the careful adaptation of established protocols or the precise development of new methods, taking into account factors such as the tissue, media composition and culture conditions. Central to this optimization of culture conditions is the achievement of stable and efficient transgene integration and expression. Sometimes these approaches are insufficient to provide reliable protocols and alternative strategies based on the application of genomic methods to genetic mapping are being investigated.

Genetic mapping: quantitative trait loci (QTL)

In plants, agronomic traits such as yield and disèase resistance are complex because they are often controlled by several genes (i.e. polygenic), each segregating according to Mendelian laws. In addition, these traits are also influenced by environmental factors. These traits are referred to as quantitative traits and the loci controlling these traits are known as quantitative trait loci.

Considering their importance, the improvement of quantitative traits is among the primary objectives of many plant-breeding programmes. In order to identify generation lines that contain the QTL alleles that contribute to the trait under selection, molecular markers associated with individual characters are identified in segregating populations and used to construct linkage maps that dissect the quantitative trait into single Mendelian components. This linkage data is used to estimate the number of loci controlling the genetic variation, to map their position in the genome and to elucidate the environmental influence on gene action.

Various types of markers are used to map QTLs. Originally isoenzymes were used as molecular markers, but recently more precise molecular DNA markers such as RFLP, RAPD, AFLP and sequence repeats (SR) of DNA sequences including minisateflites (10-45 bp) and microsatellites (2-6 bp). RAPD, AFLP and SR techniques are based on PCR and are quicker than RFLP techniques. Once the associations of markers with QTLs have been defined, phenotypes can be predicted based on the absence or presence of markers. In addition, mapping techniques can be used to identify genes for a particular character, which can be cloned, sequenced and transferred to other species.

Since the inheritance of the capacities for callus formation and for plant regeneration is complex, QTL for these characters that are constant over a range of culture conditions can be used to screen and select suitable germplasm for transformation, culture and regeneration. This technique is known as marker-assisted selection (MAS). In some crops, for example maize and rice, anther culturability is a quantitative trait controlled by nuclear encoded genes. QTL information for this trait should help in selecting genotypes that are culture responsive and to transfer the regenerability trait to genotypes that respond poorly. In rice, this would aid haploid breeding in *indica* cultivars that display low plant regeneration from anther culture.

Production of Pharmaceutical Proteins in Plant Cell and Organ Cultures

The idea that plants could be genetically engineered to produce recombinant proteins on a commercial scale for animal and human diagnostics and therapeutics is not new. Hiatt et al. (1989) were the first to demonstrate the expression of functional antibodies in tobacco leaves. Since then the technology for the production of biopharmaceuticals in plants has progressed and it is now recognized that plant systems are a cost-effective alternative to microbial and animal cell cultures for the production of proteins such as enzymes, antibodies and vaccines. Currently, there is a high demand for cheap sources of human therapeutic proteins like erythropoietin (used to treat anaemia) and insulin (used to treat diabetes). In the future, this demand is likely to be extended to a new generation of protein-pharmaceuticals that will be developed by exploiting information obtained from the human genome sequence. Plants are potentially a cheap source of recombinant pharmaceuticals, and several biotechnology companies are now patenting plant expression systems for drug production and actively conducting field trials. Moreover, a few of the plant-derived products have already made it into the early stages of human clinical trials. A number of plant expression systems are being developed for the production of biopharmaceuticals.

There are several economic and technical advantages that favour the use of plant systems as potential large-scale sources of these recombinant products:

1. Low risk of product contamination by mammalian-specific pathogens;
2. Correct polypeptide folding and multimeric assembly;
3. The relative ease of introducing new transgenes by sexual crossing;
4. Avoidance of ethical problems associated with the use of animals.

Plant tissue culture productian or whole plant agriculture?

Like whole plants, plant cell and organ cultures such as hairy roots that have been transformed with mammalian genes are capable of producing a wide range of recombinant proteins. Three main methods for the commercial production of recombinant proteins in plants are being explored:

1. Large-scale agricultural production by field-grown crops;
2. Agricultural production using greenhouse-cultivated plants;
3. Plant cell and organ cultures grown in bioreactors.

Several factors need to be considered in deciding which plant system is appropriate for production. In economic terms, field-grown production is by far the most cost effective. In some cases where the product will be delivered in the edible parts of plants (e.g. edible vaccines) without the need for extraction and purification, the downstream processing costs are greatly reduced. However, environmental concerns and regulatory issues arising from the open-field cultivation of transgenic crops could be formidable obstacles to this method of production.

Some of the concerns like environmental contamination can be met through rigorous containment by using greenhouse cultivation. Nonetheless, recombinant protein production in greenhouse-cultivated plants is more expensive than in field-grown crops by an order of magnitude.

At present, although plant cell and organ cultures are unlikely to be cost competitive with field-grown and greenhouse-grown plants for the largescale production of foreign proteins, cell and organ cultures have a number of inherent technical advantages that could be beneficial for production. For example, cell and tissue cultures grown in bioreactors are more amenable to manipulation of growth; and protein production than cultivated plants. Development of culture, regimes to significantly improve protein expression and accumulation would therefore increase the economic viability of plant cell cultures. Another advantage of cell cultures is that their time-scale for growth and production is much shorter than that for whole plants, days or weeks compared with months. This means that supply can more easily be matched to clinical demands. In addition, the environmental risks associated with transgenic plants are avoided by growing the transformed plant cells in bioreactors. Product purification and processing is greatly facilitated in plant cell culture expression systems where the product is secreted to the medium. Protein can be separated from the cells or tissue by a non-destructive step.

Based on these considerations, it has been suggested that plant cell cultures grown in bioreactors may fill a production niche for high-value, high-purity, speciality therapeutic proteins.

The targeting problem

It has already been noted that many high-value plant products are deposited in the vacuole after synthesis and are not secreted to the medium. This has a number of implications:

1. Continuous culture from which product is harvested from the medium is not possible;
2. Cells must be lysed and extracted after harvest;
3. Product may be degraded by the acidic contents of the vacuole or degraded during extraction by mixing with enzymes of the cell contents.

Engineering cells in which products are secreted to the medium would therefore greatly enhance the usefulness of cell plant cultures. To date, such a fundamental understanding of the plant secretory pathway is lacking. While great advances are being made in understanding traffic and targeting of products, development of cell lines with altered secretion should be a major research goal. Generating inducible cell lines would be even more valuable as secretion to the medium could be initiated only when the product was being generated in quantity.

The sugar problem

Many recombinant human therapeutic proteins like antibodies and blood proteins have been successfully expressed in transgenic plants. In most cases, the plant-produced proteins are identical to those formed in mammalian cells with respect to amino acid sequence, conformation and biological activity. However, these proteins are *N*-glycosylated glycoproteins and the final steps of *N*-glycan processing in plants and mammals differ, resulting in the formation of proteins with organism-specific glycans. Although these differences may not affect the activity of the protein, other clinically important properties such as antigenicity and rate of blood clearance may be significantly affected. In a recent comparative study of the N-glycosylation of a monoclonal antibody (Guy's 13 IgG 1) expressed in mouse and transgenic tobacco, although core glycans were similar, the plant protein had complex N-glycans with (1,2)-xylose and (1,3)-fucose residues. The plant also lacked terminal sialylation. The plant specific (1,2)-xylose and (1,3)-fucose residues are highly immunogenic in mammals and are key epitopes responsible for allergenic responses in humans. Lack of terminal sialylation is known to reduce the *in vivo* half-life of the proteins.

Strategies to avoid these plant glycosylation problems include producing transgenic plants lacking specific enzymes in the glycosylation pathway or alternatively modifyingg the protein after extraction.

The problem of multiple gene expression

Most of the high-value products obtained from plants are the result of complex metabolic pathways and are often at the termini of such pathways. Metabolic control analysis has also revealed that achieving significant alteration in the activity of metabolism requires more than the alteration in activity of single genes. To date, most successful genetic modifications have essentially involved only single gene changes. Modification of the activity of several genes, or introducing a series of genes required, for instance for synthesis of a secondary product would represent a major advance and open up numerous possibilities both for factory-scale *in vitro* systems, where a pathway could be engineered into a cell type which is amenable to culture, and for field production. Apart from the experimental difficulty of multiple transformation, such a multi-gene system would require appropriate co-regulated promoters for each gene to ensure appropriate levels of activity.

Future Prospects

Colinearity and synteny

Information from the completed genome sequence of two model species, *Arabidopsis* for dicots and *Oryza* (rice) for monocots, together with the increasing availability of expressed sequence tags (ESTs) for other species, will make it easier by comparative genomics to locate genes in other species. Genes with related function are grouped together in related species (properties described as colinearity and synteny). Thus, the fact that genes are located together in one species can be used to locate similar gene groupings in another species. For instance, colinearity of syntenic loci is well conserved in the family *Podcea*. Genes identified and located in rice can then be used to identify and locate homologous genes in distant relatives of the family like sugarcane.

The future development of these strategies to harness traits like *in vitro* culturabiliry or to exploit plant genes for molecular farming will make plant tissue culture and associated techniques more effective research tools as well as widen their applicability.

Bioinformatics

The combined fields of genomics, proteomics and metabolomics are rapidly yielding large amounts of information about the genes, proteins and pathways of plant productivity. Understanding these elements permits a new level of rational design in the modification of cell culture for the production of high-value products. The combination of genome data with germplasm data will also advance plant breeding, both of conventional and genetically modified crops. Interpreting the volumes of information generated by these approaches will require both computational and interpretive expertise, and the role of the bioinformatician will become increasingly important.

Concluding Remarks

Plant cell and tissue culture is a fundamental tool in pure and applied plant science. The last four decades have seen rapid and exciting advances equal to any seen in biology and the applications of the technology have had global implications. In some areas, the application of transgenic technology has met with consumer resistance and environmental caution; techniques viewed with concern in some nations are finding rapid acceptance in others. In other areas, like in the use of cultures in bioreactors, apart from in a few instances, difficulties and costs of application have restricted their use. There is no doubt, however, that the combination of genetic modification, and cell and tissue culture applied with due caution presents immense opportunity for progress. Such progress can be expected in the improvement of production by factory-scale contained systems and in the field, both with genetically modified and non-genetically modified, clonally propagated plants. At the heart of such advance must be an improved understanding of plant biology—an understanding in part to be gained by the application of plant cell and tissue culture itself.

INDEX